Love Triangle

Love Triangle

The Life-changing Magic of Trigonometry

Matt Parker

ALLEN LANE
an imprint of
PENGUIN BOOKS

ALLEN LANE

UK | USA | Canada | Ireland | Australia
India | New Zealand | South Africa

Allen Lane is part of the Penguin Random House group of companies
whose addresses can be found at global.penguinrandomhouse.com.

First published by Allen Lane 2024

001

Copyright © Matt Parker, 2024

The moral right of the author has been asserted

Set in 13.5/16 pt by Garamond MT Std
Typeset by Jouve (UK), Milton Keynes
Printed and bound in Great Britain by Clays Ltd, Elcograf S.p.A.

The authorized representative in the EEA is Penguin Random House Ireland,
Morrison Chambers, 32 Nassau Street, Dublin D02 YH68

A CIP catalogue record for this book is available from the British Library

ISBN: 978–0–241–50569–4
PAPERBACK ISBN: 978–0–241–50570–0

Dedicated to my parents, Brad and Judy Parker,
who taught me to love triangles

Contents

Zero
INTRODUCTION

I n February 2021 a driver stood before the Supreme Court of South Australia, trying to argue their way out of a speeding ticket. Back in March 2019 they had been caught driving their Mitsubishi Magna at 68km/hour in a 60km/hour zone. They had contested the speeding charge in a lower court, where they lost their case, but had since appealed that decision all the way to the state supreme court.

Their argument? That they had put bigger rims on their car, increasing the diameter of the wheel and rendering their speedometer inaccurate. Amazing.

'I said, Pythagoras and physics, a small circle spins faster than a big circle, what evidence do I need, it's mathematics,' they explained to the court. When asked why they did not have any calculations from an expert witness, they declared, 'I'm not going to pay eight grand for an expert to come.' Which makes me feel like I'm undercharging for my opinions as a maths expert (I tend to give them away for free, unsolicited).

And in my opinion their defence did have potential. In one sense, yes, 'a small circle spins faster than a big circle' when it comes to rolling. The distance around a small wheel is, well, smaller, so it needs to rotate more times to roll the same distance as a bigger wheel. If the Mitsubishi Magna's speedometer uses the rotation of the axle to deduce the car's current speed, it would give the wrong speeds if calibrated for a different size of wheel. All of which would require calculations to back it up.

Calculations the defendant did not have. It does not sound like they argued their case very well at all. The defendant was also a serial speeder, having contested twelve speeding penalties for various reasons over the previous five years. They lost the case.

What I find most interesting, though, is the use of the word 'Pythagoras' in the defendant's argument. The situation, in fact, had nothing to do with Pythagoras. Pythagoras was an ancient Greek philosopher and mathematician who was famously all about triangles, not circles. It seems this chancer wanted to sound more mathsy while trying to fast-talk their way out of a speeding fine, evoking the mysterious 'Pythagoras' as a catch-all mathematical deity.

My theory is that Pythagoras's Theorem is the most advanced mathematics which almost everyone is forced to learn at school. As a result Pythagoras has become a kind of mascot for complicated but kind-of-pointless maths. You know something is burnt into the popular psyche when it is referenced in episodes of both *Inspector Morse* and *Family Guy*.

I think it's a shame that being bored by Pythagoras is most people's lasting impression of triangles. I love triangles! We all rely on triangles to keep our modern world ticking along. I would argue (and have done, hence the book you're holding) that triangles unlock some of the most important bits of knowledge ever discovered by humans. Triangles are the gateway to the worlds of geometry and trigonometry. Triangles help us out in our day-to-day lives and enable the civilization around us. Also, I just think they're neat.

Many people leave triangles behind once they've finished the compulsory lessons on Pythagoras, geometry and trigonometry only to be abruptly reacquainted once they start their careers. Obviously some life choices will keep you in close contact with triangles. My own experience of being a maths teacher and maths author is one example. Others are far less obvious. Whenever some kind of 'isn't maths silly' thread pops up on social media, there will be a chorus of people contradicting the narrative with stories of their own necessary maths. My favourites include an oilfield driller declaring, 'Geometry began and ended my day.' And a machinist chipping in with, 'I use trigonometry literally every day.'

The oilfield worker even spelled out how unexpectedly vital mathematical knowledge was to their career. 'One of the things I quickly learned was that your math skills, or aptitude rather, determines how far you can progress in this industry.' It was through learning geometry that they were able to be promoted from derrickhand (the person who loads the next section of drill pipe) to being the driller in charge of the operation.

Humans have also been using triangles to build the world around us for a very long time. I live just outside London, which was once the ancient Roman city of Londinium. Sometime in the first century BCE, the Romans decided to build a road joining Londinium to a town near the south coast of England called Noviomagus, known today as Chichester. Romans famously built straight roads, requiring a solid knowledge of surveying and geometry, but in this case that would not be possible.

Between London and Chichester are the Surrey Hills, featuring one impressive stretch called the North Downs which is as close as England gets to an imposing mountain range.

0.034899

It's also where I live, which makes walking and cycling scenic but exhausting. While the Romans did not know it would be the future home of me, they did know that the North Downs, and subsequent South Downs, would be too hilly to easily build roads. Even if they did get a straight road through the hills, it would be too steep for any traffic to actually use. So they decided against the straight path between London and Chichester, instead turning their engineering and trigonometric ability to a roundabout route of multiple straight sections which became known as Stane Street.

To see where the straight-line road would have gone, I used Google Maps to draw a direct line between modern London Bridge (site of the first permanent bridge over the River Thames, built by the Romans) and where the east gate of Noviomagus would have been in Chichester. The map told me it was a distance of 88.6 kilometres (55 miles) and, as my gaze drifted down the line, I noticed it came into perfect alignment with a section of the modern A3 road. The A3 turned off into south London, but then my virtual line seemed to be directly on top of a straight section of the A24. Right through the map of modern London, major roads aligned with the straight line between London Bridge and Chichester.

It is not uncommon for current roads and highways in England to follow the path of ancient streets, and suspiciously straight roads are a sure sign the route is Roman. The fingerprint of Roman engineering lives on in the modern road system. These fragments of the A3 and A24 are fossilized remnants of where the Romans put Stane Street. They started building the road directly towards Chichester for the first 20 kilometres before deviating to the east to avoid the North Downs. What startled me, though, is that the road never picks up that line again. The straight sections go around and through what is now Dorking before arriving at the east side of Chichester, but the road never again aligns with the straight-line path.

I couldn't believe it. The Romans had completed the geometry required to make sure a road goes directly towards its 88-kilometre-away destination, but only for the first 20 kilometres. This would require surveying the entire straight-line distance (over two hill ranges) just to make sure the first quarter of the road matched. It would have been a shorter, more direct road if they had aimed straight for the pass through the North Downs, but no, they aimed for Chichester first, spending untold human time, effort and calculations, measuring triangles across difficult terrain, just to show off. Or, as I choose to believe, to celebrate the wonder of using triangles to survey and measure distances and angles.

If you know where to look, you can see signs of the triangles – and geometry in general – which make our lives possible just about everywhere. Mostly they work invisibly behind the scenes thanks to a host of skilled, maths-savvy experts, but every now and then we mortals can see evidence of the secret world of triangles. Sometimes that's because, like the Romans, someone was showing off.

I believe it is time that more people knew of the wonders of triangles, complete with the geometry and trigonometry triangles make possible. This blasé attitude of not caring about shapes cannot go on! Look at this packet of biscuits. An obviously eight-sided biscuit is labelled as 'hexagonal'. An eight-sided shape is an octagon! Can you imagine the manufacturers of this packaging allowing a similarly egregious mistake to slip in, like a spelling error or a mislabelled ingredient?

How about a company called Octagon Timber Flooring whose logo is not an octagon but a 3D shape, the icosahedron.

That is a bit hard to stomach.

Would you trust these people to measure a room for flooring?

I'm a big fan of an icosahedron – it's made from twenty triangles after all – but it's not an octagon. All we need now is a third product or company which is called Icosahedron but is actually a hexagon and we'll close this cycle of chaos.

These other shapes matter just as much as triangles. As I will argue, all shapes boil down to a collection of triangles but, more than that, these situations reveal how surprisingly relaxed we are as a society when it comes to geometric precision. I think it's a symptom of people viewing geometry as one of the inconsequential things they had to learn at school, the details of which they are now free to forget. With this book, those days will hopefully soon be over.

We already have plenty of people on board, both triangle enjoyers and those who use triangles professionally. Some people who love triangles really love triangles. I know of at least two of my friends who have a tattoo of a triangle. As well as the practical we are surrounded by fun triangles: the omnipresent 'play button' symbol is a triangle, the best instrument in an orchestra is the triangle, the ultimate shape to cut a sandwich into is the triangle. Triangles are great!

As I was writing this book under the working title *Love Triangle*, the fantastic stand-up comedian James Acaster released his Netflix special *Repertoire: Recognise* containing the joke 'every triangle is a love triangle when you love triangles'. I couldn't agree more. I want everyone to love triangles. And to which mathematician did Acaster attribute this fictitious maths catchphrase? Pythagoras, of course.

So let's plot a path through the world of geometry and trigonometry to celebrate all things triangle. Not a straight-line path, but one that picks out a more convenient route through North Down passes and places full of dorks. Wherever you may be on the glad-to-never-do-geometry-again to love-triangle spectrum, I hope I can show you the useful, the vital and the pointless aspects of triangles.

Triangles are everything and everything is triangles.

One

GOING THE DISTANCE

It's the first time in twenty-odd years of practice that I've had to use a maths expert and it might not be the last.

— Lawyer for the pigs

From a distance, hot-air balloons are a serene form of air transport. Slowly and silently drifting through the sky and looking colourful while they're at it. Up close, though, they are angry, out-of-control gas barbeques attached to thousands of square metres of fabric and a life-critical picnic basket. The safest way to enjoy a hot-air balloon ride is from the ground, a fair distance away from it. Specifically, half a kilometre away.

Because the sound of the burners being ignited in a hot-air balloon is so loud and terrifying, the UK Civil Aviation Authority has restrictions on where a balloon can fly, so as to avoid farms where the animals could be spooked by a screaming sky orb. In defiance of these no-float zones, in April 2012 a balloon from Go Ballooning flew over Low Moor Farm in North Yorkshire, ignited its burners and caused a pig stampede.

I'll save you the gristly details but suffice to say the farm

0.139173

lost a lot of pigs that day, including pregnant sows. Because they would now have 800 fewer piglets that season, the farmers decided to sue the balloonists for damages. Go Ballooning insisted their GPS data shows the balloon never dipped below an altitude of 750 metres. The pile of dead pigs suggested it came much closer. And, while the farmers had no proof, they did have a prof.

The best non-pig evidence the farmers could offer was a photo taken by one of their neighbours, who was far enough away for the hot-air balloon to be back in the 'serene' category. From their vantage point the balloon looked quite pleasant (the pig carnage was hidden behind some trees), which is why they took a quick snap. Along with that photo, the farmers had the power of maths. And not just any maths: triangles.

The farmers' lawyer reached out to the mathematics department at the University of York and Professor Chris Fewster took the case. Fewster's actual area of research is not calculating balloon altitudes but rather quantum field theory and the curvature of spacetime. Which, it could be argued, qualifies him to measure balloons up to several light-years in size. At a minimum, he was capable of wielding some triangles. When I asked Chris about it, he said calculating the altitude of the balloon involved 'little more than trigonometry and some understanding of how a camera works'.

Fewster could work out how high the balloon was from a single photo because triangles are 'nature's sudokus', just waiting to be solved. In fact, they are more like a series of interlinked sudokus where each one provides some clues for the next. All Fewster had to do was find a series of triangles in the photo and get solving. Sometimes the only difference between doing geometry for a practical purpose and doing it for fun is the context of the triangles.

Pigonometry

Fewster had some information about the photo: he knew where it had been taken and how big the balloon was. There were also some trees visible in the photo, which he knew the location of and, after returning to the scene of the flying, had measured with a golf laser range-finder (it appears golf has become more high-tech than I appreciated) so he also knew how tall they were. These measurements were like the clues already in the sudoku, and thanks to triangles he could now fill in all of the missing numbers.

The superpower of triangles – what makes them useful here and in so many other practical applications – is that they are easy to decode. Every triangle is made of three sides and three angles. And if you know as few as half of those measurements, you can swiftly calculate the rest. Know only the lengths of all three sides of a triangle? No problem: you can quickly use trigonometry to find the exact size of all three angles. You can only measure one side and two angles? Boom. You've got all other side-lengths and angle sizes without lifting a finger (except to use a calculator). Actually, I'd like to update my analogy. Triangles are like a super-easy crossword where you get half the letters and the answer is always 'triangle'.

I used exactly the same type of triangle trickery when NASA released this photo of the Moon in front of the Earth in 2015. It was taken by the Deep Space Climate Observatory satellite which had the Sun behind it and was looking directly at the Moon in front of the Earth. It's a fascinating photo, showing the alien 'dark side of the Moon' completely lit by the Sun with our familiar Earth in the background. I think it made most people reflect on how the far side of the Moon is called 'dark' because we can never see it from our point of view on Earth, not because the Sun never illuminates it. My first thought, purely because I wanted a reason to try and work it out, was 'I wonder how far away that spacecraft is from Earth?'

If you measure the relative sizes of the Earth and Moon in this photo, the Moon comes out to be 36.6 per cent of the width of the Earth. Which is too big! In reality, the Moon is only 27.2 per cent as wide as the Earth. The photo makes the Moon look bigger than it is because the Moon is closer to the camera than the Earth. Which makes sense; on average the Moon keeps a healthy 384,400-kilometre distance away from the Earth.

I calculated that for the Moon to look that specific amount bigger, it would have to be 74.3 per cent of the way from the spacecraft to the Earth. I knew the rest of the distance from the Moon to the Earth was 384,400 kilometres, and so it took a mild bit of algebra to calculate the total spacecraft–Earth distance. Which I think we all know I did. Details below if you want to double-check my working out.

0.190809

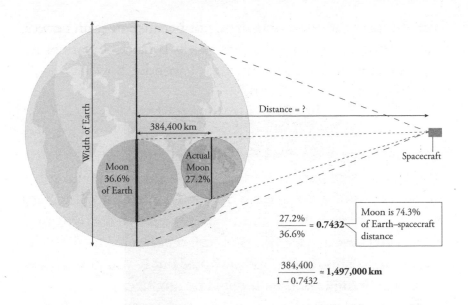

My crude calculations gave me a distance of roughly 1.5 million kilometres, which is around 930,000 miles. The NASA press release provided with the image said the spacecraft had taken the photo 'from a million miles away'. It seems both NASA and I are giving approximate answers. And even though my calculation has ignored all details of the space camera's actual optics in favour of some quickly sketched triangles, I dare say I'm closer to the true answer.

Chris Fewster did not have the luxury of ignoring the optics of the camera which took the hot-air balloon photo. He even factored in the angle the camera would have been pointing when taking the photo. Fewster was able to solve all of the triangles and calculated that the hot-air balloon was between 750 and 760 metres from the camera, which put it around 300 metres from the pigs. Much closer than the balloon operators had argued it was. They were, in the English vernacular, 'telling porkies'. Thanks to Fewster's mathematical expertise, the farmers won a decent settlement. When

I asked him about the whole incident, Fewster said he was pleased to have given 'a good example of how even basic maths can make a difference'.

The One Annoying Step

The only hitch with solving triangles is that you need to know at least one side-length. Earlier, when I said you could solve all the values of a triangle (all three sides and all three angles) with 'as few' as half of the measurements, I was covering myself because at least one of those measurements needs to be a length. This may sound obvious, but if you don't know how big a triangle is, then it could be any size.

Two triangles can have all of the same angles, yet be vastly different sizes. Imagine watching a triangle coming towards you. Hopefully a friendly one. As you wonder what you did to attract the triangle's attention, you may also notice that, even though it is looking bigger and bigger, the angles are always the same. It's like how if you walk away from a clock the hands may look smaller but the time doesn't change (well, depending on how fast you walk). The point being, if you're using triangles to calculate a distance, all the angles in the world will not help you unless you also measure at least one side of at least one triangle.

When my friend Hannah Fry and I needed to know how tall the Shard is (the tallest building in London, right by London Bridge), we did it with triangles. Yes, we could have just looked it up, but that would defeat the purpose. We wanted to recreate one of the earliest ways of measuring the size of the Earth, which used a mountain. London is sadly bereft of mountains so we swapped in a skyscraper and, true to the exercise, we calculated its height ourselves. Mathematics can so easily become clinical calculations on a page but we wanted the experience of doing the literal legwork.

I had hand-crafted my own giant protractor, which we used to measure the angle from the ground to the top of the building from two different locations. Which would all be for nothing if we didn't also know a distance. Luckily, I'd also made attachments for my shoes which were exactly 50 centimetres and stepped out 100 metres between our two angle measurements of 22.9° and 20°. Some quick number crunching later and we had a height of 263 metres, which is reasonably close to the official viewing-deck height of 244 metres. Not bad for a home-made protractor and carefully calibrated clown shoes.

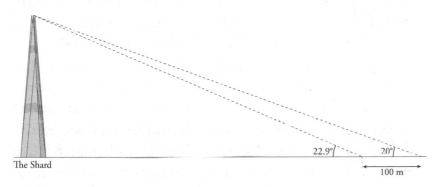

The Shard

22.9°   20°
                    100 m

100 m = 200 clown-shoe steps.

But awkwardly taking exactly-50-centimetre steps while a friend pretends they don't know you is not the only option. On a visit to Japan, I learned that Tokyo had the tallest tower in the world: the Tokyo Skytree. Tokyo is very keen to specify 'tower' because there are two buildings taller than it: the Burj Khalifa in the UAE and the Merdeka 118 in Malaysia. Apparently, buildings can be used for residential or commercial purposes whereas a tower is as useless as it is skinny.

The Tokyo Skytree is the tallest pure tower. But, you're wondering, how tall is it? I was too. So I measured it with a ruler.

If two triangles have identical angles, we say they are 'similar'. This just means they are the same triangle but scaled up or down. Hence if you know the length of any side in any one of them, and the scale between them, you can solve for all lengths. So I picked up a map of Tokyo. What is a map, if not a scale model of the city it represents? The best physical map I could find had a scale of 1 to 20,000. Which means that if I could find a pair of similar triangles – one on the map and one in real life – I could measure the map one with a ruler and know that every 1 map millimetre would represent 20 real-world metres.

I set off across the streets of Tokyo, dodging through crowds and wandering down backstreets, trying to find the end of the shadow cast by the SkyTree. Eventually the top of the shadow passed over what seemed to be a viewing-deck area next to a train line, and I was able to set my map down at its exact terminal point. I took out my ruler and stood it up on the map at the exact point where the real-world tower stood. To be fair, I've just used the word 'exact' twice, which I fear may be over-selling my accuracy. It was a partly cloudy day in a very built-up city but I was doing the best I could.

I was now looking at two shadows: one made by the actual tower, cast onto the streets of Tokyo, the other made by a ruler on a scale map of the same city. Importantly, both shadows were being made by the same sun. The Earth's Sun. And my ruler was at right angles to the ground, just like the tower tends to be at all non-Godzilla points in time. That meant the triangle made by the tower and its shadow was a similar triangle to the one made by my ruler and its shadow, which ended at the same point on the map.

As the ruler was transparent, it was casting a projection of the measurement notches on the map. I had a look to find the one which was right on top of where the actual shadow ended. 28 millimetres. I now knew that if a 28-millimetre-tall mini-tower was placed on my 1-to-20,000 map, it would cast exactly the same shadow on Map Tokyo as the SkyTree did on Real Tokyo. Multiplying 28 millimetres by 20,000 gave a height of 560 metres. Now that is a tall tower! Over half a kilometre high! Sure, the tower is actually 634 metres high and I was off by 74 metres, but for eyeballing a low-detail tourist map on uneven ground I'm very happy being within 12 per cent of the true height. If I'd just decided the ruler shadow was an extra 3.7 millimetres I would have been spot on!

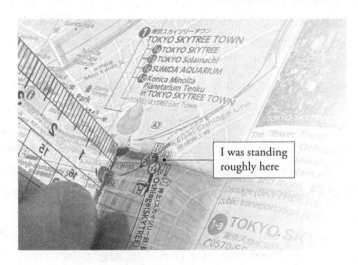

Even on holidays I try to stay very measured.

I'm far from the first person to use a shadow to measure the height of something. I'm not even the first person to use a shadow to measure the height of the world's tallest tower. In the 500s BCE there was a Greek individual named Thales of Miletus who apparently used shadows to measure the height of the world's tallest tower (while on vacation,

I assume). In Thales's case he was in Egypt visiting the Great Pyramid of Giza.

There are conflicting reports of how Thales went about this measurement. Some versions of the story say that he waited until the exact moment in the day when his shadow was equal to his height. This is the moment when the solar shadows hit the ground at exactly 45°, meaning that anything standing straight up would form a right-angle isosceles triangle, where the length of the shadow is exactly equal to the height of the object. At that magical time of day you can measure the shadow of anything and it will give you the height of the object.

The difficulty is that you need to be there at just the right time for it to work. Other writers describe Thales doing exactly what I did: measuring the shadow of a stick standing at the tip of the pyramid's shadow. But Thales could not just walk into the nearest tourist centre and ask for a free map like I did. He had to also measure the length of the pyramid's shadow to get the scale factor. But once he'd done that extra step his calculation would have been identical to mine.

The important point is, when I spend time running around cities with a ruler and a map I'm continuing an ancient tradition of mathematicians on holiday, and not, as my friends and family claim, 'wasting my vacation time' or 'confusing the locals'.

Triangles of the Ancients

The pyramids are old. As those stones were being dragged into place, woolly mammoths were still roaming around hoping these new pesky humans were not going to be a problem. When Thales of Miletus decided to try measuring their height, they were already over 2,000 years old. And it would have taken some serious maths to build them; humans have been using geometry for a very long time.

One of the earliest mathematical texts is a papyrus from Egypt, and sure enough it has a bunch of triangles slap bang in the middle. Sometime around 1550 BCE a scribe named Ahmes made a copy of an even older document from a few centuries earlier. That original document is long lost. And the handful of other, older maths documents we do still have are largely anonymous, making Ahmes arguably the earliest ever named maths author. Which I feel a profound connection with because every few years, for a tiny window of time after I publish a book, I once again become the latest ever named maths author.

The Ahmes Papyrus is now kept in the British Museum. It's believed it was originally nicked from a ruined building near the Ramesseum (temple of Ramesses II, much like I call my house the Mattesseum), but that is impossible to confirm. At some point the papyrus pinchers cut it into two pieces of 3 metres and 2 metres long, potentially to increase the resale value of now-multiple papyri. It was sold in 1858* and eventually donated to the British Museum.

Because the publication of this book would mean that Ahmes and I would once again bookend all of human known-authorship mathematical literature, I asked the British Museum if I could visit the papyrus. Due to the detrimental effects of light on papyri it is rarely put on public display, but they very kindly brought it out of its darkened storage room so I could take a gander. The first thing which struck me was that it was clearly a mathematical text. I could see triangles all over it.

---

* It was bought by Scottish lawyer A. H. Rhind, and so is often called the Rhind Papyrus. An 18-centimetre chunk from the middle was sold separately, some of which turned up in the New York Historical Society's collection in 1922, but the rest is lost.

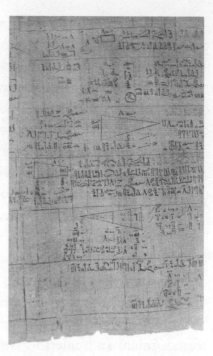

Unmistakably triangley.

The Ahmes Papyrus is basically an ancient textbook that presents a series of maths problems and then shows the calculation tricks required to solve them. The triangles which had first caught my eye are a series of problems about calculating the slopes of various pyramids, which does feel a bit too clichéd to be true. But, in a real sense, looking at the problems which were deemed important enough to go into a textbook gives us an insight into what that society was like. Within limits, of course. Imagine what historians thousands of years from now will make of our modern 'Mary is buying seventeen watermelons and two hats'-style problems. Twenty-first-century textbooks will give the impression we did a lot of shopping and bought silly things.

0.325568

The Ahmes Papyrus contains around eighty-eight different problems along with guides to solving those problems. As always, the ancient obsession with bread and grain shines through: ten of the problems are about how to divide a number of loaves of bread between a certain number of people. Six examples show how to calculate the volume of grain in variously shaped granaries. But then things take a turn for the much more geometrical. As well as the pyramid calculations, there are six problems that involve calculating the areas of plots of land. Clearly there had been some discussions about how to fairly reallocate land, presumably after the Nile had flooded and removed the previous dividing markers.

Ancient Egyptians thrived because of the fertile food-growing soil deposited by the routine flooding of the river Nile. This obviously motivated them to develop some advanced astronomical and calendar maths to predict when this was going to happen each year (leading to the direct ancestors of our modern calendar), but they also needed to know how to re-divide the land after the waters cleared. Here we see the birth of geometry from the need to calculate the exact size of plots of land.

And this is not just my speculation: the ancient Greek historian Herodotus proposed the same idea when writing *The Histories* in 430 BCE. He also added the extra flavour of suggesting that Egypt gave the world geometry because the Nile flooded each year, whereas Babylon (which was a team effort with the earlier Sumerian civilization) produced all the other maths things.

> And any man who was robbed by the river of part of his land could come to Sesostris and declare what had happened; then the king would send men to look into it and calculate the part by which the land was diminished, so

that thereafter it should pay in proportion to the tax originally imposed. From this, in my opinion, the Greeks learned the art of measuring land. The sunclock and the sundial, and the twelve divisions of the day, came to Hellas from Babylonia and not from Egypt[.]

  – Herodotus, *The Histories*, Book II, paragraph 109

So millennia ago people were already writing about the ancient human use of maths to solve arguments. And to sort out tax problems, it seems. In this situation the farmer wants to pay the minimum amount of tax but the king wants to maximize; hence everyone is motivated to get the maths exactly right.

It was humbling to look at the Ahmes Papyrus and realize it was a remnant from 4,000 years ago, when a need to divide land up fairly led to the birth of geometry as an area of human knowledge. I was particularly amazed to see a problem involving the area of a circular field which required a rudimentary value of pi. I followed the calculation and they were using an effective value of pi approximately equal to $4 \times (8/9)^2 = 3.16$, which is close to the true value. But more importantly these hieroglyphics were the gateway to all the amazing abstract mathematics involving our friend pi that humans would go on to find. I snapped a quick selfie with the maths problem, much to the amusement of the nearby historians.

From France to the Galaxy

Forget large fields and the tallest towers, let's skip right to the end and find the biggest thing we can possibly measure with triangles. As well as our instinct to look at big things and want to know exactly how big they are, humans love looking up at

An old maths artefact pointing at a papyrus.

Inside that box is a sketch of the circular field; everything below is numbers.

the night sky. For millennia we must have also been wondering how far away the stars are, with no earthly way to complete that measurement. We're going to do it. We're going to use triangles to calculate exactly where we humans sit in the universe. We're going to measure something astronomically big.

0.374607

Right at the top of the 'big things in space' hierarchy, which, as a quick reminder, goes:

- We live in a solar system of one star* and several planets, plus some other rocks and dust.
- A bunch of stars all together form a galaxy.
- A bunch of galaxies makes a cluster.
- Put a group of clusters together and you've got a supercluster.
- Those superclusters are then arranged in what is called the Cosmic Web.

If we could identify and name parts of the structure of the Cosmic Web they would be the biggest things our current scientific understanding allows us to measure. You can think of the Cosmic Web as a bit like the terrain of the Earth, with all its ups and downs. Within the chaos of that terrain there are aspects which are clearly all part of the 'same thing', and so we give them a name. Like 'the Grand Canyon', 'Mount Everest' or 'that hill I regret cycling up'. The Earth's surface may be continuous, but we understand these sections as cohesive entities we can name and measure.

The Cosmic Web is more complicated than the surface of our planet: it is a 3D foam of galaxy superclusters bigger than we can comprehend. But it's out there. Or rather, it is all around us. When you look up at the night (or, indeed, day) sky it is right there, expanding out in every direction. Except it's of a size so far beyond us that there is no immediate local structure to see. You could walk the Nullarbor for a lifetime and never understand the shape of Australia. For that, you need to get a fair distance away and look back.

* By mass our solar system is over 99 per cent the Sun, so in all honesty we're just a rounding error to our local star.

The structures within the Cosmic Web are so big that only sections which are a long way away can fit within the viewing field of our humble human telescopes. And that is a loooo . . . oooong way away ('long' with twenty-six o's). The problem with being that far away is that so little of the light from those superclusters reaches Earth that we cannot detect them. The only thing we can see at such a distance is something called a gamma-ray burst. Second only to the Biggest of Bangs, a gamma-ray burst is the most energetic event in the universe. We're talking a massive star supernova-ing itself to become a black hole. Or two neutron stars getting to know each other a bit too well. We're actually not sure: no two gamma-ray bursts appear to be the same. 'Gamma-ray burst' is more of a catch-all term for when something of unimaginable energy occurs and a burst of gamma rays (very high-energy photons) blasts across the universe.

Which appears as a tiny blip on our detectors. Often a quick blip: one in three gamma-ray bursts are over in about two seconds. Combined with how hard gamma rays are to 'focus', they are very hard to study. They were only discovered in the 1960s, by accident, when countries started investing in nuclear-explosion-detecting technology. Even in a secret nuclear test, the reaction will send gamma rays every which way right across and through the Earth, and so finely calibrated gamma-ray sensors were developed to detect these tests from countries away. But then they spotted faint spots of gamma rays coming from space.

'Faint' is relative, of course: they are actually extremely energetic, just a long way away. And they are also quite rare. Turns out it's a once-per-galaxy, every-couple-hundred-thousand-years-or-so situation. Which is good for us: a gamma-ray-burst event anywhere near Earth would be less a tiny space-flash and more an extinction-level disaster. But because there are something like 100 billion galaxies within eyesight of Earth, these events happen fairly often across the universe as a whole.

To actually study these fleeting death-blips, humans needed a way to detect the gamma rays and immediately point a telescope in the source direction to see what off-Earth was going on over there. And, believe me, a space scientist has never met a problem they think a spacecraft cannot fix. So in 2004 the Swift Gamma-Ray Burst Explorer was launched, a NASA effort with hardware from around the world. Fun fact: the UK space lab where my wife works built one of the optical detectors.* As soon as a burst was detected, the goal was to have the spacecraft swing around to point in the direction the extreme-energy gamma-ray photons had arrived from. While it couldn't spin around within the two seconds, it could get there fast enough to spot any lower-energy photon afterglow.

A bunch of interesting science things were obviously discovered blah blah blah. But that is a different story for a different book (one written by my vastly more qualified wife). What we care about is the statistical distribution of those gamma-ray events. And for the most part they were randomly arranged in the sky, basically indicating that whichever way you look, there are a bunch of galaxies over there.

Which is not to say they were completely evenly distributed. Random does not equal even. If you start flipping a coin and it perfectly alternates between heads and tails you wouldn't see that as random. Quite the opposite. And the gamma-ray

---

* Which means my research for this section mainly involved opening my study door and yelling my questions into the house. True to her profession, my wife's most common response was to suggest I launch myself into space.

distribution had exactly the amount of clumping you would expect from a nice homogeneous universe. Except for a few cohorts of bursts which were all so close together – statistically significantly closer together in all three dimensions – that there must be an above-average number of galaxies grouped together as some kind of galaxy superstructure.

The first candidate for 'biggest thing in the universe' is called the Great Wall, which, unlike its terrestrial counterpart, can be seen from space. This is a clustering of nineteen suspiciously close gamma-ray bursts, implying an area with more than the usual number of galaxies all hanging out together. Which makes for a chunky-enough part of the Cosmic Web to get its own special name. It's 10 billion light-years end to end, about 4 billion billion times longer than Earth's puny Great Wall. If, that is, it exists: astrophysicists are still arguing over the statistics. And you would not believe what their solution is to solving the argument (hint: it's due to launch in 2032).

The biggest unanimously agreed upon big-thing candidate is the Giant Ring. Nine gamma-ray bursts were found all clumped into the same ring region, something that only has a two-in-a-million chance of happening randomly, so there must be a giant ring of galaxies there. Well, I say 'giant ring'. The actual name given to it, when it was discovered in 2015, was 'a giant ring-like structure', and that 'ring-like' is in there as a disclaimer that, according to the scientists who discovered it, 'evidence suggests that this feature is the projection of a shell onto the plane of the sky'. It's a giant hollow space ball. A cosmic balloon of outrageous proportions.

From our vantage point on Earth it has an angular size of 34.5° and is 9.1 billion light-years away. Or, to switch from light-years to a more terrestrial unit, 860 septillion metres. Fill in the triangle and we know how big it is (or, at least, we know how far apart these specific gamma-ray bursts were). We can do this with any object in the sky. We can point a protractor at it to measure the angle from one end to the other, combine that in a triangle with how far away it is, and we have its size. Easy. But not so fast.

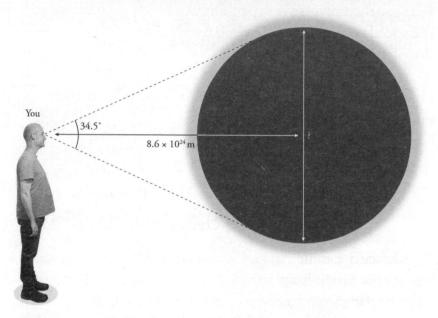

34.5°

8.6 × 10²⁴ m

You

Forget 'not to scale'. This diagram makes a mockery of the very concept of scale. It may be the most not-to-scale diagram ever produced.

How did we know it is 860 septillion metres away? How do we know how far away anything in space is? It's too far to physically measure so we need to calculate it somehow. But surely the answer cannot be another triangle because that just bumps the problem down one step: how do we know how big that triangle is? Another triangle? From here we start to pull on a thread that just keeps coming, like a retractable measuring tape. Brace yourself as we descend a ladder of ever smaller triangles until we hopefully land on something solid.

0.453990

The Redshift Rung

The distance to the giant cosmic balloon is calculated using 'redshift', which means the light coming from it is closer to the red end of the spectrum than it should be. This redshift is the result of a sort of doppler-effect where the movement of an object leaves a fingerprint in the light it produces. Because the universe is expanding at a nicely predictable rate, the speed something is receding from us is a very good measure of how far away it is.

Astronomers meticulously documented a bunch of objects where they knew both their redshift and their distance from us to deduce the relationship between the two. This resulted in a conversion factor called the 'Hubble constant' that turns redshift into distance. But how did they know the distance to those reference objects?

The Standard-Candle Rung

A 'standard candle' is any astronomical object where we know how bright it actually is. There are some stars that fluctuate in brightness at a rate dependent on their absolute brightness; some supernovae always go bang with the same brightness. When we look at these stars and supernovae in the sky, however, they appear to be different levels of brightness. We can use our knowledge of how bright they actually are to deduce how far away they must be. But we still need some absolute distances to calibrate this scale.

The Parallax Rung

With parallax we get to some real triangles. This is the effect where, if you move your point of view around, objects seem to change their alignment. Which can be very insightful. For example, many copies of the *Mona Lisa* exist, some of which were produced in Leonardo da Vinci's own studio by his fellow painters. But when one 'copy' of the *Mona Lisa* was cleaned in 2012 the conservators noticed that her hands, face and clothes all lined up slightly differently. This parallax effect meant they could tell that this was not a mere duplication of the original as previously thought, but rather was painted by someone else at the same time as the original. Because her nose lines up slightly differently with the rest of her face (as well as many other similar alignments) researchers calculated that this painting was made by someone else in the same room, standing slightly to the left of Leonardo and about 1 metre closer to the model.

In theory, if we are moving relative to the stars, we should see them shift about due to parallax. That is what sci-fi movies are trying to show when a spacecraft accelerates and stars start to whiz by. Except in real life the stars are so far away that it looks much more boring. NASA's New Horizons spacecraft is the only case of star movement being noticeable from a moving vehicle. Launched in 2006, it passed Pluto in 2015 and then, in 2020, at a distance travelled from Earth of over 4 billion miles, scientists decided to see if the stars looked any different. They turned the camera to the two closest stars (Proxima Centauri and Wolf 359) and compared them to the view from Earth. Behold: actual stars racing by a spacecraft over the course of 14 years.

For distances travelled less than 4 billion miles the parallax of the stars is simply too small for humans to notice with our pathetic naked eyes. But with telescopes and precise scientific equipment it is possible to track the very slight movement of the closest stars over the course of a year. If the same 'close-ish' star is observed at opposite points in the Earth's orbit we will see it move ever so slightly relative to much

The view from Earth and the view from 4 billion miles away.
Look at Proxima Centauri fly by!

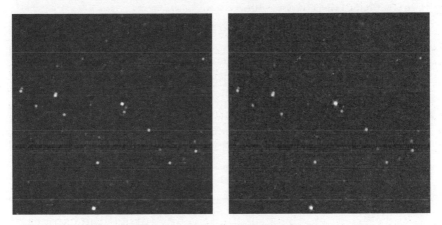

Likewise, Wolf 359 is shooting past (it's the one in the middle,
racing off to the left).

more distant stars. Which allows us to deduce how far away
that star is.

The Transit-of-Venus Rung

This parallax method requires knowing one side of the triangle: how far the Earth is from the Sun. For a long time we
simply did not know. Astronomers cheated by declaring a

0.500000

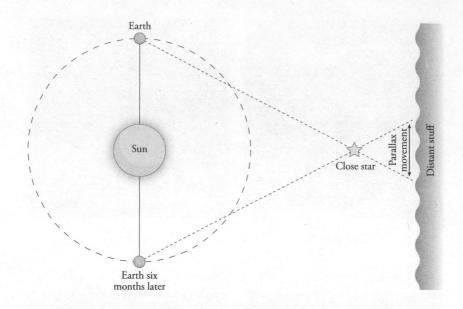

new unit, the 'astronomical unit', which just so happened to be exactly the distance from the Earth to the Sun. Even now, you see distances to stars stated in astronomical units. But that is not so much measuring the distance as just labelling it.

How can we get an actual value for the astronomical unit? You guessed it: more parallax. More triangles. It's an understatement to say that the Sun is super bright, so parallax is hard to measure. Except for when something moves between the Earth and the Sun. Thankfully, Venus passes between the Earth and the Sun a couple of times every century or so. If this is observed from multiple locations on Earth it will look like Venus takes slightly different paths and that can be used to get the distance to the Sun . . . if you know the size of the Earth.

The French-Countryside Rung

The first modern calculation of the size of the Earth was thanks to two French mathematicians in the 1700s. Jean-Baptiste Delambre and Pierre Méchain spent the better part of

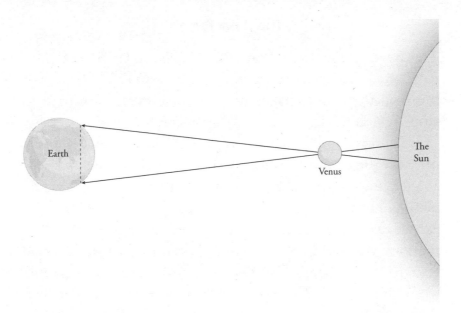

a decade mapping out a chain of 115 giant triangles across 1,500 kilometres, from Dunkirk in France down to Barcelona in Spain. It was not an insignificant task. Delambre and Méchain started with one giant triangle with each corner on a hill so they could see the other corners and measure all of the angles. Then they marked out a second triangle which shared an edge with the first one, and then another, and so on, each triangle contacting one of the previous ones. All they did was measure angles because measuring angles is easy. A triangle may be miles across, but the angles will still be small enough to measure with a protractor (albeit a fancy, accurate one, made of brass).

At each end of this chain of triangles they knew the exact latitude: a measure of how far from the equator Dunkirk and Barcelona were (this is why they had deliberately run their triangles in a north–south orientation). They used the latitudes to deduce what percentage of the complete circumference of the Earth they had managed to cover. If they then calculated the length of the chain of triangles, they could scale-up that distance to give them the total size of the planet.

The Ruler Rung

This is where we step off the size-calculation ladder and onto solid ground. We've been kicking the measurement can down the road for several pages now. But eventually you have to stop kicking the can and start measuring the road.

Delambre and Méchain had to get down and actually measure the length of one side of one of their triangles with a ruler. The 'ruler' in question was four carefully calibrated platinum rods, each paired with a strip of copper because the metals expand at different rates in the heat and so any fluctuations with temperature could be adjusted for. After all four rods had been carefully laid end-to-end, the rear one would be moved to the front. Ruler after ruler they inched their way down the road. Actually two roads. This task fell to Delambre, who spent forty-one days measuring a road just outside Paris and then a later forty-three days measuring a second road in the very south of France. These two roads had been included as edges in the mesh of triangles because they were fantastically straight. And they measured two sides so they had one to double-check the other; hence the phrase, 'measure twice, calculate 115 triangles once'.

Once they had confirmed these edge-lengths they could calculate every side of every triangle, one after the other, until they knew the exact distance from Dunkirk to Barcelona. And therefore the size of the Earth. And therefore the distance to the Sun and the distance to standard candles. Which allowed for a calculation of the Hubble constant to turn redshift into distances. The gamma-ray bursts in the Giant Ring had redshifts of between 0.78 and 0.86, which we can now calculate and average out to our 860 septillion metres.

At last we can confirm that this behemoth space bubble is truly massive. From side to side it is 5.6 billion light-years.

A phone call across the shell would take 5.6 billion years each way. The universe is only 13.7 billion years old, so the whole call so far would be 'Hello how are you?' followed by 'Not bad, you?' and then nothing.

The starry sky-orb is actually a substantial chunk of the sky, we just can't see it because it's so faint. If it was bright enough to see it would appear about sixty-six times wider than the Moon. And if it were patterned like a hot-air balloon this is what the night sky would look like:

That's no moon.

Some astronomers argue that something this big is not 'gravitationally bound', which means the disparate sides are so far apart they are not linked together by gravity in the way the Earth and Moon are, or a piece of toast and the ground. So there is an argument that it's not really a 'thing' as such. At some point we could just consider the entire universe as a thing we can try to measure. But I'm not letting go of my space ball. We can argue semantics but it's definitely a large structure within our observable universe. The biggest we've found yet. And we can measure it with a triangle. It doesn't matter how big the balloon is you want to measure: triangles have got you covered.

I have picked one series of rungs from the cosmic balloon to the ground. There are plenty of other techniques for estimating distances and sizes in the universe, but all of those alternative ladders also use triangles. I imagine these days astronomers can just bounce a laser off the Moon or something.

But whatever the case, my point stands. The reason Delambre and Méchain were trying to measure the Earth was because the new-fangled 'metre' had just been defined as one 10-millionth of the distance from the North Pole to the equator. Their measurements are what gave us the metre and all other metric distances. To this day, if you take a step of 1 metre you have walked one 40-millionth of the distance around the planet.

So whether you are measuring the size of the biggest ball in the Cosmic Web or the height of a balloon above a pig, it is all possible because in the 1700s two humans got out a stick and measured the distance of a road in France.

Two

A NEW ANGLE

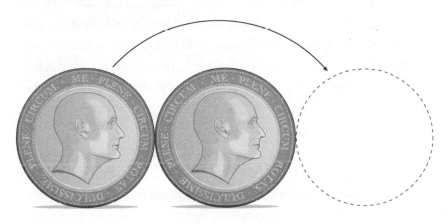

If you roll one coin around a second coin, which way up will it be? Spoiler below.

This is a famous coin puzzle. See if you can keep track of which way up the first coin will be as it rotates around the second one. Turns out that handsome head will not be too dizzy as the coin will end up back the same way it started. This throws people, as it feels like she should go through a half rotation and be upside down. Keeping track of angles in our heads is not always quick and easy. But it's worth it because of what angles can open up.

Look around the world, you'll see angles everywhere. Much

like distances, angles can give us an insight into how the world around us works. When the same angles keep popping up we know there is some kind of logic going on below the surface. The wake behind a duck on a pond always forms an angle of 39°. Big duck, small duck; fast duck, slow duck: always 39°. Which tells us something about the way waves move in water. When a doodlebug digs its sand trap to catch ants it builds walls at 34°. Which is the exact same angle as the leading edge of a sand dune. That 34° tells us something about the nature of sand. A rainbow is always 84° across. Which tells us something about the nature of magic and friendship.

But what is an angle? I'd say an angle is just a difference in direction. Like, when you say goodbye to a friend you want to make sure there is a difference in the direction you are both about to be walking, otherwise it gets real awkward. You either make small talk until you say goodbye again, or you come to a silent arrangement to pretend you don't know each other. In this case the angle between your directions is 0°.

The best case is that you walk in exactly opposite directions, the biggest angle, the full 180°. There are a bunch of other angles in between including the one where you move in a perpendicular direction, 90°, the threshold angle between acceptable and awkward.

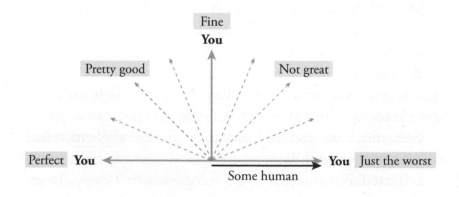

You may have noticed I've already had to resort to that little floating circle above numbers which is pronounced 'degrees' (though you're welcome to pronounce it 'oooOOOoo'). As in degrees of full circle. In a perfect world all angles would just be measured as a fraction of the biggest angle, but no. Way back in prehistory some genius decided that a full rotation should be split into 360 degrees. A lot of people blame the Sumerians and their love of 60 (360 being six lots of 60) and others have pointed out that the Earth's orbit is roughly 360 days, so yeah, that. Whatever the case, we're now lumped with 180° being the biggest angle. There are other units angles can be measured in like radians and gradians, but much like the rest of society we are going to ignore them (for now).

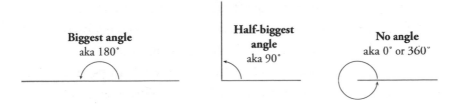

Biggest angle
aka 180°

Half-biggest angle
aka 90°

No angle
aka 0° or 360°

I call 180° the 'biggest angle' but that is not strictly correct. True, any angle bigger than 180° can be replaced with the smaller angle 'from the other side', but I'd argue that is subtly different. Angles up to 180° give you a measurement, whereas bigger angles tell a story because they say something about the history or context of the angle. And angles 360° and bigger still have a role to play. They keep track of how many times something has rolled. The 360° the coin in the puzzle rotates through is very different to the 'equivalent' 0°.

Sometimes we want to know the complete angle travelled. If you turn around twice you will find yourself back where you started. But you've also gone somewhere. Physically and emotionally. So you might want to say you've been through 720°, even though your current position is identical to 0°. And nowhere is this more obvious than the world of skateboarding.

In skateboarding a 720° is literally twice as rad as a 360°. For the uninitiated, when skateboarders talk about a number of degrees that is a measure of how much they, and the skateboard beneath their feet, have rotated while in the air on a single jump. In 1999 the professional skateboarder Tony Hawk was the first person to land a '900°'. While in the air he and the skateboard rotated two and half times (360° × 2.5 = 900°) and then successfully landed (the landing part is quite important). If you looked away while he was in the air, it would appear like he had just turned around, rotating a mere 180°. The first ever '1080°', which occurred in 2020, would look even more underwhelming.

I love a sport which has embraced measuring rotations using degrees with such enthusiasm. But I can confirm, via painful teenage experiences, a knowledge of angles does not itself help with being proficient on a skateboard. In skateboarding, angles are used as a kind of bookkeeping to track the awesomeness of a trick. For my knowledge of angles to give me an advantage in a sport, we need a game that gives you time to do some geometric planning. Thankfully there was one sport where I could pool my knowledge.

I Never Felt So Good

Angles were responsible for my only significant sporting achievement in high school, on a PE outing to a pool hall. Now was my time to shine. During my uncharacteristic display of sporting success, a fellow student expressed surprise at how well I was doing. Almost immediately a (much more sporty) friend of mine piped up with 'well, what did you expect, pool is basically all angles'. My maths reputation being as well-established as my lack of sporting prowess.

It's hard to find a more common or more practical use of angles and geometry than playing pool (and snooker, billiards, the whole ball-on-felt family). It must be the most direct interaction with angles that most people have in a recreational setting.

There is much to be said about the mechanics of balls colliding with each other, but we'll leave that mess to the physicists and just look at when a ball hits a cushion. In theory, this should just follow the rule that 'angle of incidence equals angle of reflection'. Like light in a mirror. Which is exactly what you'd expect intuitively when an object ricochets off a surface. A grazing shot with a small angle of incidence is going to reflect only slightly and carry on in a similar direction. A shot going straight at a surface is going to bounce right back at you. At least, that's the theory.

According to that theory, a mathematician should be the ultimate pool player. To test this, I formed a team with my mathematical friend Grant Sanderson and we challenged two non-mathematical professional pool players to a shoot off. They set up a practice drill shot where the cue ball had to be bounced off two specific cushions and then hit a target ball at the far end of the table. Team Maths got to work straight away, putting the cue stick down and whipping out a tape measure. Once we'd measured everything there was to measure we retired to the bar to do some angle calculations. The pool players entertained themselves playing a game on the adjacent table while taking breaks to heckle us.

We took the mirror analogy of 'angle of incidence = angle of reflection' quite literally and, instead of calculating the angle the ball would need to bounce back on, we

0.642788

imagined it rolling straight onto a second, phantom, flipped table. When you look in a mirror your brain does not think you're looking back in the reflected direction; rather, it seems that there is a second copy of reality in front of you. And mathematically it all works out. You can try this yourself with a mirror and a laser pointer. If you want to bounce the laser off a mirror and hit a specific target, you can either work out the exact angle of incidence you need, or just point the laser at the target as it appears in the mirror's reflection. The maths works out the same way in either situation. And it works for as many mirrors as you have lying around near your laser. In our case, we needed two phantom, mirror-world pool tables.

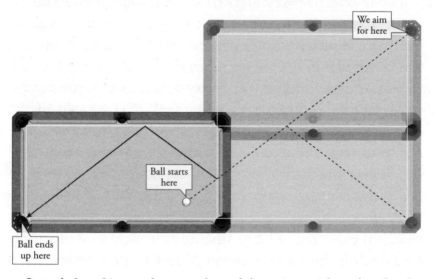

Instead of working out the two angles needed, we aim straight at the reflected phantom pocket.

We then set a target at the exact location in the room where the target ball would be located two fictitious tables away. Which was a cup we balanced on a stool. But we knew that if we aimed for it as if it was a real pocket in the distance, the angles would take care of the rest. So we set it up and . . . missed horribly.

This did not surprise the professional players in the slightest. They knew something we didn't: the more a ball bounces around the table, the more the angles deviate from their theoretical, perfect trajectories. But not in a random way; the angles changed in a logical way the players had come to anticipate. They talked about shots 'opening up', and we soon realized this was because when a pool ball bounces off a cushion it causes the ball to spin and that spin alters the angle of any future collisions. It's the story of my life: foiled by reality not conforming to my rough approximation of it.

On a perfectly frictionless table, or one where the balls are point masses (that is, balls unencumbered by having a physical size and instead moving as an ideal hypothetical particle), our maths would have worked out very nicely, thank you very much. And, in reality, pool players do what we were attempting except they know how to compensate for the spin effects. They showed me some 'calibration shots' that they do as a standard test whenever they are playing on a new table to check how the friction of the felt and the compression of the cushion change the behaviour of the ball.

Professional players actually use reflections in more ways than we expected. They sometimes line up a shot in one location on the table, pick an object in the distance, and then use that as a phantom pocket to aim for in future shots. They may not realize they are aiming at a pocket on a hypothetical mirror table; they just know that it works. They also explained that you can aim for the pocket on the next table over in a pool hall, and the ball will go into a pocket on your table. It works because the tables are roughly one table-width apart!

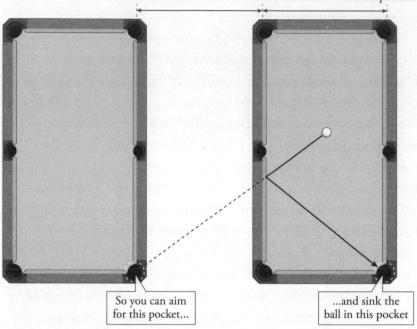

So you can aim for this pocket...

...and sink the ball in this pocket

But what if a table did ever reach true frictionless nirvana? This leads to the interesting hypothetical question: on a perfect pool table is it even possible to miss? I seem to be an attractor for strange questions, and this one was posed to me many years ago by someone who followed my work online and it caught my attention. Specifically, if a ball bounces around a table without stopping will it inevitably drop into a pocket or could it technically continue to bounce around indefinitely? There is a trivial solution where the ball just bounces back and forth at right angles to the cushions. Which does answer the question: yes, it is possible to miss. But it seemed boring to stop there. To see if there were more interesting answers I used phantom tables to 'unwrap' the reflection trajectory of an infinite pool shot.

Instead of one finite table, I imagined an infinite sea of

pool tables – each one a reflected copy of its neighbours – tiling out infinitely in every direction to make a never-ending 2D surface of pool tables. The question of 'can a shot miss?' becomes 'is there a direction where a ball can travel out in a straight line and avoid the infinitely many pockets it will pass?' There are two solutions. One is what I did at the time: find a path where the ball ends up back in the starting location, heading in exactly in the same direction.

I made some pool-table graph paper and drew a line which started and ended at the cue-ball spot without going through any pockets. Job done. The ball can follow that straight path forever. But that does feel slightly like it's just a more convoluted version of the back-and-forth bounce shot. After every 'two tables down and six across' the ball is back exactly where it started. Translated back to a single table, it would bounce around off four cushions, looping along that exact same path forever.

The more interesting solution would be a path which never repeats. And I felt like that should be possible. If the pockets are all on 'whole number' integer coordinates on this

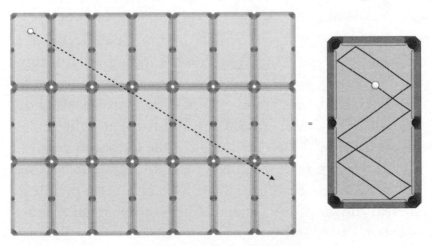

Now that's my kind of infinity pool.

0.694658

pool-table grid, then hitting the ball on an irrational angle should mean it never hits an integer set of coordinates. But this assumes the pockets have effectively zero width. In reality, the pocket has some width and that messes with the maths (and even I feel like infinitely small pockets is bordering on cheating). At this point I fear some readers have tuned out, while some will be tempted to try and find a proof of my snooker conjecture! I admit that we have ended up far from the practical world of sports, but it just shows how many recreational ways there are to enjoy angles.

There is No End of the Rainbow

It seems to be a tradition in my books to pay good money to a stock image website just so I can be angry at one of their pictures. This image is advertised as a 'stock photo' and described as 'Double rainbow after the rain in the sky over the field'. See if you can spot the four things which make me so angry. I'll give you a hint: it's not a real photo.

Lazy work. Somewhere a designer was totally over rainbows.

We'll reflect on the double rainbow in a moment, but the first glaring problem is that the rainbows are totally the wrong shape. Most people, if asked what shape a rainbow is, will probably say it's an arch. Which is not correct. They are so close, though. All rainbows are actually perfectly circular, and the maths name for part of a circle is an 'arc', so they're off by one letter. In this case, the photo fraudster has stretched out the rainbow into more of an oval shape. Which simply cannot happen.

The reason for the circleness is the very process which causes rainbows to form in the first place: the refraction of light. This is similar to reflection but doesn't automatically send light back on at the angle it arrived on. Rather, it bends the path of light based on the properties of whatever substance the light is going into. The infamous 'speed of light' is only technically the speed of light in a vacuum. The moment light has to actually move through something it slows down. For example, it slows down a bit in air, and even more in water. Actually, 'a bit' may be overstating air's ability to slow down light. At normal, human-friendly temperatures and pressures light moves through air at 99.97 per cent the speed it would in a vacuum. Water slows light down to 75 per cent of its maximum speed.

The crazy thing is, when light changes speed it also changes direction. For complex physics reasons. This is why, if you view something through an air–water boundary, it will look bent. And the exact angle of that change depends purely on the speed change of light as it goes from one medium to the next. A rainbow is the result of light refracting in and out of spherical water droplets in the air. The light refracts into the droplet, some of it reflects off the back (much of the light does pass through, though; it's like a two-way mirror where some light reflects and some passes through) and then it refracts again out the front. This has the net result of turning the light around.

Which brings us to the next part of rainbow geometry and the next mistake in the photo: a rainbow is always directly opposite the Sun from your point of view. In fact, everyone sees a unique rainbow, and if you were to draw a line from the Sun behind you to the middle of the circle of rainbow in front of you, that line would pass exactly through the middle of your head. Technically, as you move your head around, the rainbow is following you. This is another reason the stock image must be fake, as you can see shadows on the hills in the background, implying the Sun is off to the right. Any photo of a true rainbow will have shadows which point directly away from the camera.

All of these properties of rainbows are due to how they are formed. Rainbows are closely correlated with rain because for a rainbow to occur there need to be small water droplets in the air. There also needs to be direct sunlight, which is why they are associated with the end of a bout of rain: there is still water in the air but the clouds have parted to let the Sun through. The water is absolutely key, though. Technically rainbows don't even exist; what you're looking at is light coming directly from the water droplets. What you can see *is* the water droplets. Let's take a look at the reflectors in the mist.

We'll start with the red wavelength of light. Below we have the cross section of a spherical drop of water. The light from the Sun comes in on parallel lines and, for simplicity, I'm only showing the light hitting the top half of the droplet. The light has three opportunities to change angle: refraction entering the droplet, reflection off the back of the droplet, and refraction as the light comes back out of the droplet. Those three angle changes interact in some interesting ways.

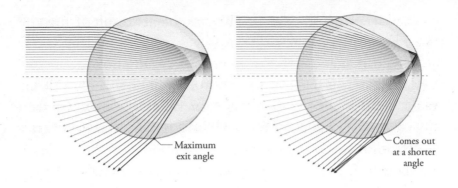

Maximum exit angle

Comes out at a shorter angle

Light which slams into the middle of the droplet barely deflects at all and comes almost straight back out with only a slight change in angle. Looking at incoming rays hitting the droplet further up from the centre, you can see they exit on bigger and bigger angles. But, interestingly, this eventually hits a maximum level of deflection. Because of the knock-on effects across the three angle changes, rays which hit the very top of the droplet actually exit with a smaller change in direction. My diagram shows all of the light paths but in reality that maximum-angle region is by far the brightest point because the droplet has kind of focused the incoming light together.

But enough about the droplet! What do we see? When the light from the Sun is coming directly from behind you, all the droplets of water which are within that maximum angle, from your point of view, will reflect some red light back at you. But the ones in the middle will be quite faint, and the ones at the edge much brighter because that is where all the rays of light are overlapping. The end result is that you will see a large, red disc in the sky with a bright edge.

Red is not the only colour, though. It is the one with the longest wavelength (aka 'big disc energy'), while all the other colours with shorter wavelengths have smaller maximum angles and so appear as smaller discs in the sky. All of these discs overlap and the light is all combined together.* Around the edge of the biggest disc you will see a bit of bright red sticking out, just inside that you'll see the bright ring around the yellow disc, all the way down to the smallest of the discs, violet, right in the middle. Inside that you have all of the colours in roughly equal measure which combine to give generic, white light.

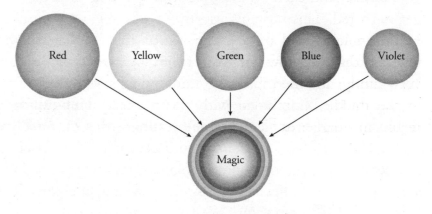

A rainbow is actually a stack of progressively smaller coloured discs.

If you are not convinced by this, having spent your whole life thinking of rainbows as arches and not as giant sky-discs, the next time you see one take a look at the middle. The area inside a rainbow will be distinctly brighter than the space just outside it. Note that, compared to the background, it's not like a rainbow disc will be blinding. In the diagram before last, I should have indicated that when the light hits the far side of the water droplet most of it passes through and out

---

* To make it easier to follow, I'm explaining everything as discrete discs of colour when in reality the spectrum of light is a spectrum. But we're all used to rainbows being approximated as a series of bands of colour, so it should be fine.

the other side, and only a small bit gets reflected back towards us. Making a rainbow less bright than having the Sun shine directly in your eyes.

I also showed the rays of light passing back out of the droplet on the third angle change when in reality a small amount of light will get reflected again within the droplet and pass out on the next bounce for a total of four angle-alterations. And each time the light is reflected the colours get mirrored. This flipping also means there is now a minimum angle instead of a maximum angle, resulting in an inverted disc. So, now, the rest of the sky gets lit up (albeit very faintly) and then there is a rainbow-edge to a dark disc in the middle. Combined with the primary rainbow, this results in a dark area between the two rainbows.

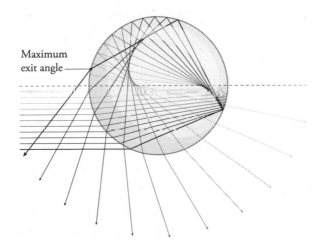

Someday we'll find it, the rainbow reflection.

With each reflection the rays get more spread out and some light is lost, so everything gets fainter. Strictly speaking, the light makes even more bounces causing a third rainbow, but by then it is so faint that humans eyes cannot see it. The point is the second rainbow has a fraction of a fraction of the original light, which is why you only see a double rainbow when the sunlight is particularly bright. And it is always fainter than the first rainbow. Hence the third mistake in Exhibit A: the second rainbow is brighter than the inside one. This cannot be the case, as the second, outer rainbow is the result of more lossy bounces within the water droplets and is by default more dim.

Which brings me to the most egregious error in the stock image. A normal, primary rainbow has red at the top and then goes through the colours down to blue at the bottom. But because it is reflected one extra time the secondary rainbow is the other way around! If you are ever lucky enough to see a clear double rainbow, you will see that the outer secondary rainbow is not just fainter, it also has blue on the top and red on the bottom. It is this one fact that means you can glance at a page of double rainbow images and immediately split them into fake and real (or rather, fake and convincing-fake/real).

For completeness, here are all the parts of a real, full-circular double rainbow as photographed from a helicopter. I appreciate that you are reading about colour spectrums in a book printed in greyscale, which does take some of the shine off a rainbow, but I've helpfully tried to label everything for you.

If you follow all of the bouncing-around of red light within the water droplet, the final angle of deflection is about 42°, which is why I said earlier that all rainbows are 84° across. Likewise, you can extend the calculations for one more reflection and determine that the minimum angle of the outer rainbow is 51°.

So always remember, when you're marvelling at a double rainbow you're actually looking at a stack of 42° coloured discs inside a second stack of 51° inverted discs. All centred

Reverse spectrum —•

•— Spectrum edge of centre disk

Slightly brighter out here

Dark region between the rainbows

Bright centre region

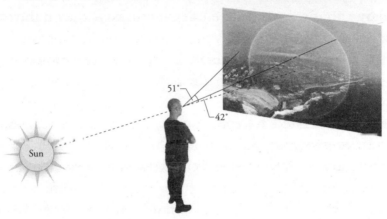

51°

42°

Sun

perfectly on your own personal Sun–head line. Now that's magic. But, ironically, a rainbow is something you cannot technically share with a friend.

Angle of Attack

Some recent research shows that the asteroid which ushered the dinosaurs off the stage impacted the Earth at the worst possible angle. From the dinosaurs' point of view. For us mammals, who took the spotlight, it could be considered the optimal angle. Arguably, humans only exist because of that one, specific angle 66 million years ago.

When an asteroid slams into a planet, obviously anything standing directly under the impact is going to have a real bad day, real fast. In this case, 100 million tonnes of bad day at a minimum of 43,200km/hour. But the rest of the planet does not necessarily get off lightly. An outrageous amount of rock, dust and just about anything else that an asteroid impacts will be ejected up into (and sometimes through) the atmosphere. A bunch of that will rain back down, but enough can hang around in the sky to block out the Sun for a substantial amount of time, which is bad news for anything that needs the Sun to grow, and therefore eventually for anything that likes to eat things.

How much material gets sent skyward depends on what angle the asteroid hits the Earth at, so in 2019 some scientists decided to work this out. They ran computer models of massive, city-sized asteroids hitting the ground at angles of 30° (fairly shallow), 45°, 60° and 90° (flying straight down). These models calculated in 3D what would happen to all of the rock unlucky enough to have the job of stopping the asteroid, revealing not only how much stuff would be blown sky high, but also what the resulting impact crater would look like.

The actual impact at what is now Chicxulub, Mexico took place about 66 million years ago in what was a fairly shallow ocean, up to about 1 kilometre deep (which sounds deep until you remember the asteroid was over 10 kilometres in diameter, so when it hit the bottom it was barely submerged at all). The millions of years which have since passed mean that what is left of the almost 200 kilometre-wide crater is long buried, covered in hundreds of metres of ocean sediment, half still under the water and half under Mexico. The impact was so powerful that not only did it wreak havoc with the Earth's crust, it also disturbed and deformed the Earth's mantle over 30 kilometres below the surface. And that sort of disturbance hangs around for a long time.

We know the crater is there thanks to magnetic scans of the area showing deformities in the rock, gravity measurements revealing the density of the material below the surface, and exploratory holes drilled down to remove rock core samples. This was all initially done for oil exploration, and it took a while for the research to be released to the scientific community who have now also sent several expeditions to check out the impact site. Geophysicists used seismic reflection and refraction data to scan the shape of whatever is down there. All of these measurements have revealed a slight asymmetry to the crater, implying the asteroid must have come in on an angle.

This is where the computer models came in. Scientists looked to see which of the impact angles produced a 3D shape in a computer that most closely matched the actual shape of the crater under Chicxulub. From this they concluded that the impact would have been quite steep, somewhere in the 45° to 60° range. Something around 60° is the worst-case scenario, as you get more material in the atmosphere than either more shallow or more steep angles (sorry dinos). As the scientists from the Department of Earth Science and Engineering at Imperial College London who did the study put it:

> A steeply-inclined impact produces a nearly symmetric distribution of ejected rock and releases more climate-changing gases per impactor mass than either a very shallow or near-vertical impact.

You can kind of picture why. A shallow impact will move less material as it's more like a glancing blow. Something coming straight down may move the dirt but might not send it into the sky. However, this sort of intuitive speculation around asteroid impact can be unreliable as they are such otherworldly events, involving energies not normally seen on Earth. You might throw some stones into dirt and look at those impact craters but that will teach you very little about an asteroid impact. For example, the vast majority of impact craters on other surfaces are circular, while stone-throwing experiments on Earth suggest they should be elliptical.

A thrown stone is moving well below the speed of sound, and uses its momentum to move material and give a crater a few times its size. A supersonic asteroid clears material via an expanding shockwave and clears out a crater 10 to 20 times its size. Because that shockwave radiates out so far from the impact site, it does not end up mattering that the asteroid would have physically scooped out an elliptical impact, the much bigger blast radius ends up being roughly circular.

Computer modelling is a fantastic tool which has helped humans understand everything from financial markets to magnetic fields in the Sun. But, no matter how fancy the model, it does need the occasional sense check to make sure the computer code simplification has not deviated too far from reality. Which necessitates some experiments to check how closely physical reality matches the mathematical pre-dictions. I think you can see the problem when it comes to asteroid impacts.

We could wait for another major asteroid impact. But it would be better to make our own impact by throwing something very, very hard. NASA decided to turn the tables by throwing something at an asteroid itself. Not only would that answer some impact questions, it would also be a test to see if humans are capable of diverting the path of another celestial body.

In 2022 NASA's DART mission (Double Asteroid Redirection Test) slammed an object just over half a tonne in mass into an asteroid at around 6km/second. This pre-emptive strike wasn't on an asteroid which posed any risk to Earth. Quite the opposite: the asteroid was picked because, while it was close to Earth, it was definitely not going to hit us no matter how much DART knocked it off its trajectory. The headlines 'NASA Redirects Asteroid into Earth' were about as bad a PR disaster as is possible. Also, while it wasn't a revenge slap for the dinosaurs delivered 66 million years after the fact, it was the first time in the billions-of-years-long asteroids-versus-Earth war that the Earth finally struck back.

Humans had landed spacecraft on other celestial objects before, and even impacted a few, but this was the first ever off-world experiment to see if we could redirect an asteroid. A skill worth learning before it is strictly needed. It cost $330 million to get the 580-kilogram impactor careening into the asteroid at 6.15km/second*, which will seem like a bargain if we ever need to do it with any level of urgency. And it's not like the DART spacecraft had to carry the impactor all the way there. The DART spacecraft *was* the impactor. Possibly the least smart spacecraft in recent history, it was a metal box with just enough sensors, cameras and propulsion to find the asteroid and slam into it. If we're sending the spacecraft equivalent of iPhones to explore Mars, this was a Nokia 3210 we lobbed at a rock.

The DART spacecraft did have one bit of clever kit: a

---

* For error-bar fans: technically it was 579.4 $\pm$ 0.7 kilograms of impactor travelling at 6.1449 $\pm$ 0.0003 km/second.

detachable camera. But even that was nothing fancy. The LICIACube was a tiny, CubeSat-style spacecraft with two means of taking images: LUKE (wide-angle colour images) and LEIA (narrow-field but high-res greyscale images), whose names are acronyms so tortured I refuse to reproduce them here. Think of the whole system as a Nokia that comes with a free polaroid camera.

NASA chose the asteroid not just because of where it was, but because of what it was: a two-body system. The main asteroid, Didymos, is 780 metres across and orbiting around it is the adorable, 160-metre Dimorphos, which is the one we hit in the face. A binary system is a good target because, instead of having to wait ages to see if a single asteroid has significantly deviated from its previous path, we can immediately see if the orbital rate of the asteroid pair around each other has changed. This is because any relative movement in a two-body orbital system will immediately change the period they take to rotate, and that is much easier to measure. This change in period can be used to work backwards and calculate the scale of the trajectory change.

That would have been quite the mission in its own right, but NASA is all about maximizing the science possible, and this was going to be a perfect experiment for what happens in an asteroid impact. So the LICIACube was sent along for the ride, and detached from the main DART about two weeks before the collision so it could move back to a safe distance. The high-res images from DART itself rightfully got a lot of public attention because they showed Dimorphos gradually getting closer and closer until the transmission suddenly stopped. But it was the less glamorous images from the LICIACube I was excited about because they were going to reveal some interesting angles.

Space scientists were very excited about the DART

project, even long before it happened. A collaboration of forty-one scientists from thirty different organizations wrote a paper titled 'After DART: Using the First Full-scale Test of a Kinetic Impactor to Inform a Future Planetary Defense Mission', which was one long mathematical daydream about all the things they could do with the data from DART once the impact had occurred. This included measuring the angle the ejecta would fly out of the crater-to-be. I loved seeing the complex diagrams made by scientists trying to think of all the angles they could potentially measure.

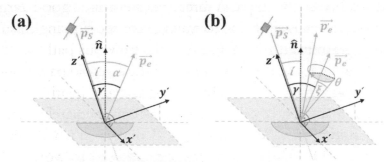

Figure 1. Coordinate system and angles. (a): The tan square in perspective indicates the asteroid surface around the point of impact, with the darker semicircle suggesting subsurface material that will be excavated and ejected. The spacecraft arrives traveling in the −z′-direction, at an impact angle *i* from the surface normal *n̂*. The net momentum of the outgoing unbound ejecta, in the homogeneous case, is coplanar with z′ and *n̂*, and directed at an ejection angle α from the normal. (b): As in (a), showing the parameterization of the random part of the response. The ejecta momentum is increased in magnitude by a factor (1 + *ζ*) and altered in direction by an angle *ξ* at azimuth *θ*.

Look at all these angles they couldn't wait to measure.

I met Sabina Raducan, one of the authors of that paper, at a space-science event my wife Lucie was helping organize. Whereas professional footballers have the old-fashioned-named WAGs (wives and girlfriends), at a physics conference there is always a group of us SAPs (spouses and partners) who are along for the ride. It gives me an opportunity to ask a bunch of maths questions of Lucie's colleagues. At this point in time, the DART collision had occurred but none of the data was public yet, so Sabina very kindly offered to show me the still-embargoed ejecta-angle paper she was working on. I'm never going to turn down secret space triangles.

In the images taken by LUKE I could make out the plume of debris flying off Dimorphos. Close analysis of those images revealed the ejected material filling a cone with an angle of somewhere between 131° and 139°. This angle was surprisingly big. A completely vertical, 90° impact should produce a cone of about 90°. This was a fairly-close-to-vertical impact (DART came in somewhere between 66° and 80° from the asteroid surface), yet it produced a much wider cone than expected. Raducan explained that, because Dimorphos was so small relative to the impact crater, the curvature of the asteroid means there is simply less material surrounding the area to contain the impact, and that caused the wider angle.

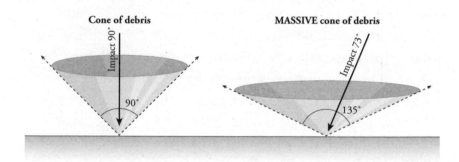

An unexpectedly large cone of debris.

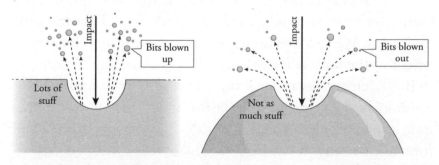

An impact on a small sphere means the thin walls can be blown away.

Sabina is a planetary scientist researching impacts in space at the University of Bern in Switzerland, and her goal was to use this angle to learn more about the composition of Dimorphos. It wasn't possible to land a robot on the asteroid to check exactly what it was made of, but the composition of the asteroid is key to understanding how the impact unfolded. This leads to a complex web of different components and factors in the collision, which Raducan needed to solve enough of to untangle the rest. This is important because if scientists don't know all of the fine details of how the impact unfolded, they cannot apply the lessons learned this time to future impacts of other asteroids.

Scientists had determined that the original time it took for the two asteroids to orbit each other was 11 hours and 55 minutes. Post-impact, this had dropped down to 11 hours and 23 minutes. The change in period was enough to calculate the exact change in trajectory of Dimorphos, but not to explain exactly why this impact had caused that exact change. The amount of momentum transferred to deflect the asteroid's trajectory depended on what the asteroid was made of, but there was more than one type of asteroid which could give the same result. As Sabina's post-impact paper says, 'multiple possible combinations of cohesion, coefficient of internal friction, and bulk density could result in the observed deflection'. So she had to make a variety of computer models to see which one matched what was seen in the sky. We can look at all the different factors she could adjust in the simulation:

Bulk density is what it sounds like: how dense the whole asteroid is on average. This doesn't take into account the individual particles. That is the job of cohesion, which is a measure of how well the individual parts of the asteroid hold together. For example, if there are a bunch of large clumps they would resist sliding over each other because they kind of lock together. But that doesn't take into account friction. If they did slide over each other there would be some level of internal friction resisting that movement.

Rather delightfully, to make a realistic simulation of what is basically a rubble pile in space, Sabina would start with a bunch of different-sized digital rocks floating about and then the computer code would apply simulated gravity until they collapse together to form a virtual asteroid. Once anything loose on the surface was dusted off to match what Dimorphos actually looked like, she was left with one possible combination of bulk density and cohesion to test with a simulated impact.

Friction could also be varied through some plausible values. And internal friction is something we've already met in passing: this is why sand dunes and insect sand traps form at a 34° angle. If you pour different substances out on a table into piles, they will form different angles depending on the friction between the particles. Something that does not have much internal friction will form a very shallow pile; for example, grains of wheat slide past each other pretty easily and so they form a pile of 27°. Higher friction articles like chalkdust will form steeper piles; in chalk's case, a 45° pile.

If you need to measure a 45° angle in a pinch, piles of shredded coconut, wheat flour and wet sand will all give you a reference 45°. This angle is called the 'angle of repose' and engineers have measured it for all sorts of substances. Dry sand has an angle of repose of 34°, which is why that pops up so much in nature. This angle depends only on the internal friction of the substance (and gravity). For the material in Dimorphos Sabina considered everything, from friction similar to glass beads (angle of repose of 22°) through to the type of rock found on the Moon (lunar regolith has angles between 35° and 45°).

Of all the many simulations, the one which best matched the angles seen in the observations was a bulk density of 2,200 kilograms per cubic metre, a very small cohesion force of less than 1 Pascal (you'll have to trust me that's small) and an internal coefficient of friction of 0.55, which on Earth would give an angle of repose of 29°. But the proof is in the pounding. The simulations produced images which had to be compared side-by-side with the actual photographs taken in space. As the mantra goes: only believe computer simulations insofar as they match what we see in reality.

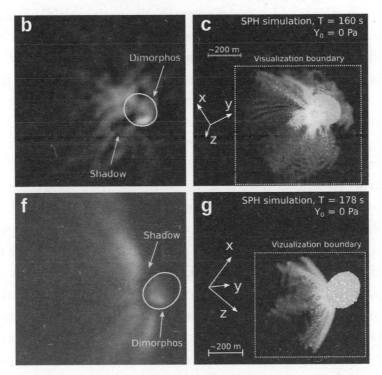

Photos to the left, simulations to the right. Here I am struck by the simulatory of the two.

This match between computer simulation and reality means that Sabina was able to report that Dimorphos was a floating pile of rubble which was barely held together. Which is important to know. The momentum change to its trajectory after the impact was actually larger than the momentum the DART impactor could have provided. This apparent paradox can be explained because all the loose material which got blasted off Dimorphos also had a momentum impact on the remaining body; it was effectively a jet slowing the main object down even more. If we do have to deflect an asteroid in the future it might be a much more cohesive object than Dimorphos, and so the collision could play out very differently.

In 2024 the European Space Agency will launch the Hera mission which is also heading to Didymos. The LICIACube was able to provide some images but, because it piggybacked a ride with DART, it arrived at a screaming pace and could only take a few frantic photos before it carried on out into open space. Hera is going to hang out with Didymos for a while, take some photos and maybe do some landing. This will give us a very detailed picture of the 'after impact' situation. Sabina predicts that 'the DART spacecraft caused the global deformation of the asteroid' and we'll not see an impact crater so much as a misshapen asteroid. Time will tell.

We still have much to learn about impact angles. Future experiments on and off the planet will help us confirm the computer simulations being developed are accurate. As the collaboration of forty-one forward-looking space scientists themselves put it, 'the DART test will serve as an initial – and, for the time being, singular – ground-truth anchor'. I look forward to many more ground-truth anchors. Except for a Chicxulub-type impact on Earth. I don't want to speak for all mammals, but one of those was exactly the right amount.

Three

LAWS AND ORDERS

C hapter Three in a book about triangles. That's pretty special. We've started to get acquainted with triangles through their sides and angles. Keeping lawyers in jobs, protecting the planet from asteroids and getting better at pool.

Now it's time to really get to know triangles. There are the visible aspects of triangles – three sides and three angles – but there are also a host of hidden laws that triangles have to abide by. 'Unwritten rules' that all triangles must obey. Well, I'm going to un-unwrite them. I've picked six of my favourite triangle laws: five of them I find delightful, and one. One of them makes me irrationally angry. So get ready for that.

Area: ½ × Base × Height

Regular, non-maths humans often have a strange sense that maths would be capable of solving all of their problems if only they knew how. Like maths is a dark art understood only by a shadowy group of experts. Every now and then a civilian, on discovering you are a mathematician, will immediately pitch a problem at you. (Which I prefer to them describing,

0.898794

in an unhealthy amount of detail, how much they disliked a maths teacher they had, which seems to be the only other option.) One of my favourite cases was a toasted-sandwich food truck operator who wanted to know how to cut a sandwich in thirds.

Halves, they explained, were easy. Wham: one diagonal cut down the middle. Unanimously agreed upon as the optimal way to cut a sandwich. That said, not strictly the mathematical optimal. A sandwich can also be cut in half perpendicular to the crust, which renders two rectangles: an equally fair division with a slightly shorter cut. But no one wants rectangles. This book is not called *Love Rectangle*. The people want triangles!

Quarters, likewise: two diagonal cuts, four small triangles. For five pieces or more, you really should be getting your own sandwich instead of ≥ five people sharing the same one. That said, as we saw on the Ahmes Papyrus, humans have been arguing about dividing up bread for at least 3,500 years. But I'm going to assume the clientele of a toasted-sandwich truck are not eating in teams of five or more.

One piece is the trivial case: just hand the sandwich over. So the missing link is three pieces. It's not unheard-of for three people to share a sandwich, or for one person to want to consume their sandwich in three stages. It is at least a popular enough division that, when a purveyor of sandwiches met a practitioner of mathematics, they demanded I explain how it could be done. And this wasn't during a late-night sandwich purchase; we were both on a TV talk-and-cooking show and this was their opening backstage chat. At the time hipster cheese toasties were all the rage, hence why a sandwich artist and a mathematics artist had both been booked.

The added complication is that people have opinions about

crusts. Sometimes they don't want them, and sometimes –
like on a grilled cheese sandwich – they are to be fought over.
So I couldn't just suggest they slice the bread into three rectangles (a maths solution with the crusts cut off); I needed
three triangles with equal sandwich and equal crust.

I grabbed a serviette and did a quick sketch of a toasted
sandwich (approximating it as a square). 'Yes!' I said. 'You
can easily divide a sandwich into three equal pieces with only
one extra cut.' Instead of cutting from one corner to the
diagonally opposite corner you make two cuts from the same
corner, each ending a third of the way along the two sides
which make the opposite corner. Which gives you two
triangle-shaped pieces and a central, kite-shaped piece, all
with the same area. But the kite has much less crust. Ah. The
mathematical purist in me wanted the ultimate solution,
where eaters of any level of fussiness would have no cause
for complaint.

'One second,' I said. New serviette. This time I started
with a cut from one corner to the exact middle of the square.
A brief pause. Yes! The problem solves itself! As long as all
cuts start in the middle, if you divide up the crust equally, all
the pieces will also be the same size. Another pause. I realized that it works perfectly for square bread and 'fails safe'
for rectangular bread, where everyone still gets the same
amount of sandwich but the crust allocation is slightly off.
The general vibe was that the solution was suspiciously easy,
with no need to measure angles. Just judge what a third of
the way along an edge is. But that is one of the many powers
of triangles.

The equation for the area of a triangle is $\frac{1}{2} \times$ base $\times$ height.
You may recall seeing $A = \frac{1}{2}bh$ in school, but the classroom
atmosphere was such that you didn't give it a second thought.
Take a second look at it now. Really look at it. It suspiciously

lacks any involvement of the angles. No angles! It turns out angles have no direct bearing on the area of a triangle. Area only depends on how wide the base of a triangle is and how high it is. And don't get hung up on which side of a triangle is the 'base': it's a free-for-all, any side can be the base.

The centre of a square piece of bread is the same 'height' above all of the sides. So if a triangle is cut from any point on the crust to join the centre, the area depends solely on the amount of crust. If you divide up the crust fairly you get equal areas, and therefore amounts of sandwich, for free! Which also cracks the case for five, six, seven and any number of people. Divide up the crust and the sandwich shall follow.

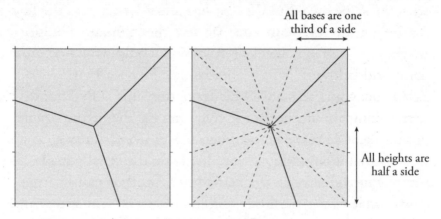

On the left is the cut-pattern and on the right the sub-triangles are also shown. You can convince yourself they all have the same height and base. Or go get a sandwich.

I knew when I started writing this book that I would want to include some everyday, practical applications of triangles. And the area of a triangle is such a fundamental property it makes a perfect candidate for pragmatic geometry. And I wasn't short of options. A lot of things have an area. Between the times I've used triangles to calculate how much paint or carpet is needed to cover an area of my house, and the occasions when I've deployed some strategic calculations to see if the odd-shaped wrapping paper I have on hand is sufficient for the gifts I have to disguise, this brief encounter from ten years ago kept coming back to me.

I wondered if my advice had had any impact in the world of sandwich cutting. So I did a bit of googling and emailed the Jabberwocky gourmet toastie van to see if they remembered me from our backstage chat. Not only did they remember me, but they replied with a photo they took of the original serviette and went on to say they have been cutting their three-way sandwiches using the Parker Method ever since. It seems my knowledge of triangles has forever changed the face of sandwiches.

Pythagoras: $a^2 = b^2 + c^2$

The granddaddy of triangle maths. First name: Pythagoras. Last name: who cares about his last name, this is Pythagoras we're talking about, the Beyoncé of maths. Pythagoras even sounds like it starts with π . . . because in Greek it does! The guy is as mathsy as it gets.

Everyone remembers being forced to learn about Pythagoras at school, though many people didn't understand why, and now here we are, with Pythagoras being a cultural touchstone for complicated maths. As well as the guest appearances in *Inspector Morse* and *Family Guy* mentioned before, *The Simpsons* also gave it a nod:

> The sum of the square roots of any two sides of an isosceles triangle is equal to the square root of the remaining side.
>
> – Homer J. Simpson
> (Season 5, Episode 10: '$pringfield')

0.927184

Which is not the correct statement of Pythagoras's Theorem. In the episode someone offscreen yells, 'That's a right triangle, you idiot!' The reason for the inaccuracy is that *The Simpsons* lifted the line verbatim from the Scarecrow in the 1939 film *The Wizard of Oz*, in which there is sadly no correction at all. D'oh.

All of that said, there is no good reason why we should call it the Pythagorean Theorem over, say, the Scarecrowean Theorem. We know very little about the ancient Greek philosopher Pythagoras, who lived two and a half millennia ago, and what we do know doesn't even confirm he came up with it. Pythagoras, unlike the scribe Ahmes, didn't have the decency to show his working out and clearly write his name on it. A popular theory is that Pythagoras popped over to Egypt, saw them using triangles and brought the idea back home. Or maybe he never went to Egypt. I'll let the historians argue these points. Either way, many other civilizations also had their own version of the theorem.

China's version is called 勾股定理 (the 'Gougu Theorem', which translates literally as the 'base-height theorem'). Pretty much everything we know about ancient Chinese mathematics comes from a handful of texts written around 2,000 years ago, plus or minus a century. What we know as Pythagoras's Theorem appears in more than one of them. I particularly like a conversation recounted in *The Gnomon of the Zhou* between a member of the Zhou-dynasty royal family, the Duke of Zhou, and a mathematician named Shang Gao (the person believed to have discovered the base-height theorem in China).

The Duke wants to know if mathematics can be used to measure the heavens, when they are out of reach. Instead of banging on about parallax and the Cosmic Web, Shang Gao decides to hit him with the base-height theorem. Which,

I must agree, is a fantastic example of how mathematics can be used to deduce the size of something without measuring it directly.

> Take the distance directly underneath the sun as the height, take the distance to the point underneath the sun as the base, multiply the numbers by themselves, then add them and take the number whose square equals the sum, and that's your distance to the sun
> — Shang Gao, from *The Gnomon of the Zhou*

It seems all civilizations, sooner or later, stumble across Pythagoras's Theorem, either by hearing about it or finding it for themselves. Trade and communication between empires did take place at this time, and the later Roman and Han empires knew of each other's existence. Or there could have been a shadowy secret society of trianglists already calling the shots back in 1000 BCE (I'm just asking questions!). Or it could be that, because triangles were so important to a developing civilization, people spent enough time looking at them they were bound to notice these sorts of relationships.

Whatever it's called, here is the theorem in a nutshell: for any right-angle triangle, if you square the length of the longest side, you get the same value as when you square each of the two smaller sides and add them together. It can be demonstrated pictorially as shown overleaf:

It may feel like a triangle law that only applies to right-angle triangles is a bit limiting. What about all those triangles which don't have a corner equal to a quarter turn? But this theorem has endured for two key reasons: non-right-angle triangles are just two right-angle triangles in a trench coat, and, actually, right-angle triangles themselves are surprisingly prevalent. Including in modern data.

0.939693

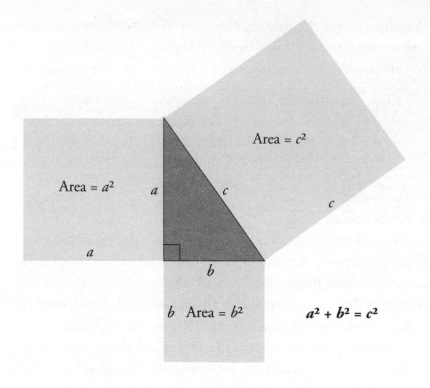

Area = c^2

Area = a^2

a

c

a

c

b

b Area = b^2

$a^2 + b^2 = c^2$

I know introducing coordinates this early in any maths lesson could be considered 'putting Descartes before the course', but coordinates drive modern data. And we can start with a nice literal example of spatial coordinates: shot data from the National Basketball Association.

The NBA has been tracking the exact x and y coordinates of every shot taken in every game since the 1997–98 season. I can look up and see that Michael Jordan's first shot that season has coordinates (–85, 199). In this case, the first number is the x coordinate, which is the short axis running parallel to the baseline, and the second number is the y coordinate, giving you the distance down the length of the court. The 'origin' that the coordinates are all relative to is the exact centre of the basketball hoop. Which means positive x values are the right half of the court and negative the left (when standing under the ring and looking out). Negative y coordinates mean the shot was taken from behind the basket (the middle of the hoop is actually 5 feet and 3 inches from the baseline so there is plenty of scope for this).

I've been watching basketball since the 1990s and it seemed to me that the average shot distance had been gradually increasing over the decades, with three-point shots becoming far more prevalent in recent years (in basketball, an ordinary basket scores two points, but a basket scored from behind the three-point line earns an extra point). To convince myself this was actually the case, I contacted my friend Tim Chartier, an expert in sports analysis and statistics who works with the NBA. He sent me back the data on all 4,678,387 shots taken during NBA games between 1997 and 2022. For each of these almost-5-million shots I now had the x and y coordinates of where they were taken from. My plot was coming together.

Interestingly, the NBA have stayed true to their American traditions by recording the distances in feet, but they've also tried to sneak in some of the conveniences of the metric system by recording the coordinates in tenths of a foot. And I am on board with 'decimal feet' as the perfect unit in this situation. So Michael Jordan's shot from position (-85, 199) is 8.5 feet to the left of the basket and 19.9 feet down the court.

My only complaint about the data was that the distance of every shot was rounded down to the nearest foot. The opening Michael Jordan shot is recorded as a 21-foot jump shot. 'That is not accurate enough for me!' I thought. Thankfully the true distance was only one application of Pythagoras's Theorem away. Or, in my case, 4,678,387 applications. So I wrote a bit of computer code to go through the entire database and add in a much more accurate distance metric for every single shot:

```
dist = int((((i[0]**2 + i[1]**2)**0.5)*12/10)+1
```

0.951057

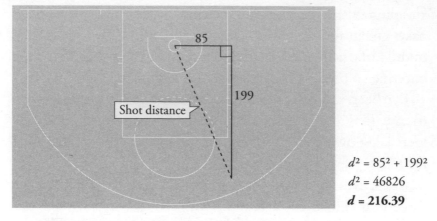

$$d^2 = 85^2 + 199^2$$
$$d^2 = 46826$$
$$\mathbf{d = 216.39}$$

Michael Jordan's shot was 21.639 feet from the ring. He missed the shot.
Pythagoras never misses.

i[0] and i[1] are the x and y coordinates. In the programming language I was using, Python, **2 is used to square something (raise it to the power of 2) and likewise **0.5 is to take the square root. The little cheeky multiplication by 12/10 is because I wanted to convert tenths-of-feet into inches, and the whole thing is wrapped in int(...)+1 as a quick way to round up to the nearest whole inch (the int function by itself rounds down, so adding one is the same as rounding up). Combined, it's not the fanciest code in the world, but it gets the job done.

Once I had the distance data I set about doing some analysis and visualizations of the average accuracy from different distances, and comparing that to the average points scored per shot. The very short answer to my original question is that yes, NBA players are shooting from further out than they used to. This is because, even though accuracy does start to drop off at long distances, the fact that players get three points for any shot beyond the three-point line, compared to two points from anywhere inside it, more than makes up for that. Instead of averaging 0.8

points per shot just inside the three-point line, players can take a step outside it and average 1.12 points per shot. That is the same points-per-shot as being 2 feet and 2 inches from the ring.

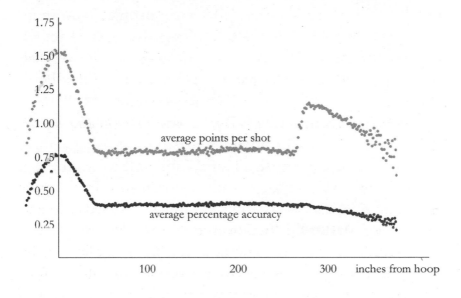

Y'all ready for maths?

I'm obviously not the first person to notice this: statisticians working for NBA teams originally crunched similar numbers and helped fuel the player move downtown. But I was the first to spot some mistakes. As I was messing around with the data I decided to label my plots using the NBA's data labels, which indicate if a shot was a three-point shot or not. Suddenly, I found I had three-point data points that appeared way too close to the hoop.

Intrigued, I dug into the data and wrote some new code to compare the Pythagoras-calculated distance of every shot to the distance of the three-point line, then cross reference that against whether the shots were labelled as two-pointer or three-pointers. I found several hundred shots where the distance I calculated did not match the shot type the NBA had recorded. There were three-point shots way too close to the ring and two-point shots from way too far out.

I sent this data back via Tim, and it eventually landed on the desk of the official NBA stats people. Who then fixed the data. The official NBA shot data was changed because of me. Normally, to contribute to the official NBA shot data during a game, people have to train hard for years and work their way up to be some of the most elite athletes in the world. But I was able to add some three-point shots to the NBA database just by using maths. All because I wrote my own Pythagoras code to crunch the distances.

Triangle Inequality: a + b ≥ c

The official name of this rule is the Triangle Inequality, which is refreshingly descriptive and not linked to an ambiguous dead person. 'Triangle inequality' doesn't refer to an unfair distribution of resources in the triangle world, but rather to what the sides of a triangle can and can't equal. In short, no two sides of any triangle can add up to less than the remaining side. There are a bunch of ways that you can prove this. Euclid did it with an isosceles triangle, but you can also just look at it. Proof by staring at it. Because if two sides combined are shorter than the third side, they simply cannot reach both ends of it.

After the unexpectedness of the Pythagorean Theorem involving squares, this is a reassuringly intuitive triangle regulation. And it's one that I learned myself through making a mistake. I distinctly remember doing some kind of mathematical activity in lower high school that involved generating a series of arbitrary triangle lengths. I wrote down a bunch of numbers to represent the three side-lengths of triangles and was shocked when I got the work back from the teacher with a note along the lines of 'these triangles are impossible'. For a triangle to be physically possible, no one side can be longer than the other two combined, because, and I don't mind saying it again, those two sides will not be able to reach each other.

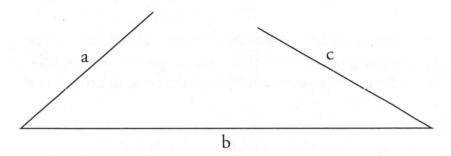

Because a + c is smaller than the length of b, no triangle is possible.

But what use can there be for something seemingly so obvious? Like all things triangle, there is always an application. For a start we can use it as a tool to see exactly how wrong Homer Simpson was. Sure, we know he didn't recite the Pythagorean Theorem, but maybe he was telling us about a different theorem that would apply to other types of triangles? Let's work it through.

> The sum of the square roots of any two sides of an isosceles triangle is equal to the square root of the remaining side.
>
> – Homer J. Simpson

Removing the requirement of 'any two sides' (which promises way more than any triangle could deliver) and stripping out the isoscelesness, we are left with the more general notion that there exists some triangle where the sum of the square roots of two sides will give the square root of the remaining side. Henceforth, the 'Homer conjecture'.

So, with three sides, 'a', 'b' and 'c', if the sum of the square roots of a and b add to give the square root of c, that can be written in algebra, as follows:

$$\sqrt{a} + \sqrt{b} = \sqrt{c}$$
$$c = (\sqrt{a} + \sqrt{b})^2$$

In that second line I've just rearranged it slightly so we get an equation for what the third side, c, looks like in terms of a and b. And if you take the time to expand out those brackets you'll get:

$$c = a + b + 2\sqrt{a}\sqrt{b}$$

Feel free to double-check that or blindly trust me. The important point is that side c is equal to the lengths of a and b combined plus a bit more. Which is in violation of the Triangle Inequality. No triangle can both conform to Homer's conjecture and be physically plausible. Which may feel like a flippant use of a triangle law, but I assure you it is just one of a host of abstract mathematical proofs and deductions which use the Triangle Inequality as a logical tool.

Just because something is obvious doesn't mean it's not worth stating clearly. In this case, the 'obviousness' that two short sides, too far apart, cannot physically reach each other belies its importance. It is obvious *because* it is so fundamental. If there is a straight line, 'c', between where you are and where you want to be, that is the shortest path. You cannot get there quicker going the long way, via 'a' and then 'b'. The

Triangle Inequality basically says that, when it comes to distances, there are no shortcuts.

Mathematicians have generalized this concept of distance in something called a 'metric space'. We already have an intuitive sense that there are different types of distances; hence we use 'as the crow flies' to distinguish straight-line distance from, say, the distance by road. But why stop there? In our physical reality we use Pythagoras's Theorem as the metric to calculate the distance between two points, exactly like I did for the NBA basketball shots. But that metric could be anything, from calculating distances around a black hole using the equations of general relativity to abstract notions of 'distance' between data points in machine learning.

Mathematicians love loosening constraints to see how generalized something can become before it loses all meaning; searching for the limit where there are so few restraints you can no longer tell the baby from the bathwater. And it turns out, for a metric set to behave nicely, there are only three rules you need to keep:

- POSITIVE: the metric gives you a positive value between any two points, and only gives zero if the points are equal.
- SYMMETRIC: the distance from a to b is the same as from b to a.
- TRIANGLE: the Triangle Inequality holds for any three points.

Even though it may feel patently obvious to say that a 'triangle law' is that all the sides of a triangle need to be able to reach each other, it is the one explicit rule we need so that even the most esoteric notions of 'distance' will make logical sense. And to prove Homer Simpson wrong.

Triangles are Rigid

I knew in general that civil engineers loved triangles, but to make sure this book was sufficiently well researched I gave my engineering mate Paul Shepherd a call to ask him how much he loves them. He was very excited to chat triangles.

> Students have this idea that 'triangles are strong', but engineers love them because they don't skew, like a rectangle becoming a parallelogram. To stop a rectangle flip-flopping about, you put in a diagonal to make it rigid. So engineers love triangulating structures.
>
> – My engineering mate

That is the whole law: triangles are rigid. This is because there is only one way to build a triangle: three side-lengths can only form one triangle (or its mirror image if you flip it over). Mathematically, we would say a triangle is completely defined by its three side-lengths. Practically, that means that for the sides of a triangle to move, the triangle itself would need to be ripped apart. If the sides are joined together well, it would take a lot of force to do that.

The same is not true for other shapes. The four sides of a rectangle can make countless diamond-shaped rhombuses just as easily. The sides of a rectangle can move around: if the corners flex just slightly, the whole shape skews to the side. Stopping corners from flexing at all is way harder than merely making sure they stay attached. Which is why the way to keep a rectangle solid is to put a cross-brace across it, which effectively turns it into two rigid triangles.

This is why engineers love triangles: if you make a triangle you do not need to worry about holding it in place or stopping the angles from moving – you get all of that for free. Whereas architects like rectangles because they look good and it is really easy to produce glass rectangles. Which is why you see so many modern buildings which are covered in rectangles.

You might think you have a concrete counterexample: bricks. And concrete! Bricks are rectangles (the only triangle bricks are ornamental). Which is a good point. The 'triangles are strong' thing largely only applies to shapes which are empty frames. Or, at least, things which can skew if not held in place. Something solid like concrete is not going to skew (it will break long before it deforms). By 'triangles' and 'rectangles' I mean the shapes that make the load-bearing structure of a building form. Which is why engineers love a good brick, but hate a rectangular window frame.

If you're unfamiliar with the life cycle of planning a building, here is my understanding of it: an architect draws a picture on a serviette, and then that gets given to a team of engineers who will spend months trying to turn the sketch into something physically plausible. I'm oversimplifying a bit, of course. I think the architect is allowed one afternoon to visit the engineers and yell at them, reiterating that windows are meant to be seen out of, and all the massive diagonal beams the engineers have added are going to block the view.

As far as I can tell, much of modern civil engineering is about sneaking triangles into buildings when the architects are not looking.

Angles Sum to 180°

If you rip the three corners off a triangle they are guaranteed to fit snugly together and form a flat line. It's a violent way to show that the three angles in a triangle always add to 180°, aka half a circle. It's not immediately obvious why that should be true. It relates to an equally unobvious fact that every journey requires rotating a full 360°. Which I can demonstrate nicely using the time I rode a MotoGP bike around the Silverstone racing circuit.

MotoGP is the extreme-acceleration pinnacle of motorsport. These bikes accelerate and decelerate more severely than even the cars in Formula One racing. I was involved in a documentary about speed, and the producers thought it would be funny to put the mathematician on one such bike and send me around England's iconic Silverstone race track at speeds exceeding 250km/hour.

Thankfully I had no direct control of the bike, as I was riding with a professional. I did have some indirect control over the bike, though, as the noticeably smaller-than-me rider pointed out that my mass was a substantial percentage of the bike's and that extra inertia would impact the handling. I took that to be exactly as insulting as they intended.

The ride itself was several shades of terrifying and was the ultimate test of how well I could hold on to two tiny handles. Thankfully, I had done some preparation. Nothing to do with fitness or speed, but I had had a good think about what data I could possibly log. As well as being on a race track, I was also keeping track. I may have been on a race circuit, but I was using some circuits. Look, I had an app.

During the ride I had an app running to record all of the data coming from my phone's built-in angle sensors. For all sorts of reasons, including trying to flip the screen orientation appropriately, phones like to know which way up they are. So they have a bunch of sensors which measure the current angle they are on sixty times per second. After the lap, I exported all of my angle data into a spreadsheet. Here is a plot of my angle relative to the direction of forward travel.

I need to make clear how messy this data was. I had strapped a phone to my arm before I was irreversibly zipped

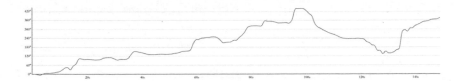

Yaw had to be there.

into a motorbike 'safety onesie' (my words, not theirs). Not only was I then sat on a motorbike which is more angry motor than bike, but my arms were far from stationary: as well as trembling in fear I had to constantly reposition my mass on the bike depending on what the greatest threat to my existence was at any point in time (accelerating, braking or cornering – the ABCs of death).

This data records turning to the right as a positive increase in the angle and left turns are negative, so every time that line goes down it means the bike was going around a corner to the left. You can see a strange, gradual increase over time. At the end of the lap, I am not back where I started, but rather my net angle has increased by 360° (within the error of me repositioning my arm). This is because I went around the circuit in a clockwise direction, resulting in a complete rotation of the bike.

You can try this much slower on a dramatically simplified racing circuit: a triangle. I'm going to call the simplified Silverstone track 'Sliverstone', and instead of a bike we're going to use a pencil. Start with the pencil on one edge of the triangle and then progressively move it from edge to edge. Each time it will rotate slightly more, until it has done a full spin and can land back on the starting line.

We can follow along with a quick bit of corner bookkeeping. At the first turn the pencil is going to rotate some angle between 0° (not turning at all) and 180° (doing a U-turn and

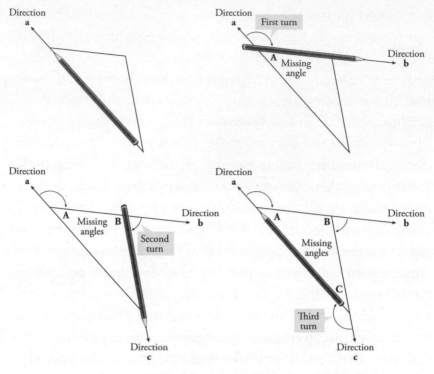

My pencil runs on leaded fuel.

heading straight back to the start). The angle inside that corner of the triangle track is the number of additional degrees needed to complete the turning angle up to a total of 180°. There are three corners during the lap, so if we were to add all three turning angles plus all three inside-the-triangle angles the total will be 3 × 180° = 540°.

The individual turns will depend on the specific triangle but we do know they must add to 360°, as the pencil needs to do a full rotation before the end. Which means the inside three angles must combine to give 540° − 360° = 180°. We have thus inadvertently proved that the sum of the angles of any triangle must be 180°. If you're excited, you can have a go at doing the same thing for a square or pentagon circuit, and show that for any n-sided polygon the interior angles add to $(n - 1) \times 180°$. If you're less excited and tuned out for some of that, don't worry: we're getting back in the saddle.

If you've tuned into the rhythm of this book, you'll know that this is when I should present some kind of fascinating application of the maths fact we've just covered. But that's a bit difficult for the sum of angles in a triangle. It's not that it is not incredibly useful; it's one of the most fundamental laws of triangles, if not *the* fundamental triangle law. But it's never really used in isolation. It's like asking what the application of garlic is: it's an amazing source of flavour, but it's not like you eat it by itself (and if you do you're going to get strange looks, probably from a great distance). You could contrive some kind of hypothetical application of it in isolation, like warding off vampires. But does that count? I hope so, because I'm about to do the triangle version of that.

Often I've found myself wondering about the angle of one thing or another, but never with such intensity as when I was on that MotoGP race bike hurtling around the Silverstone race track. You may have seen racing motorbikes going around corners and marvelled at the angle they are able to tilt over on. Well, now I've done that but from the inside of the angle. And getting that angle wrong while on a bike is much like getting garlic wrong with vampires: someone is about to lose a lot of blood.

I learned many things on that ride about myself as a person. One was that I had completely underestimated what friction was capable of. The speed you can go around a corner is limited by the amount of friction you have with the road, and this bike had far more friction than I thought was allowed by the laws of physics. To utilize that friction we had to lean into the corners so much I felt like we were trying to lie down on the road surface.

Gravity always points straight down and is famously not open to negotiation.

As we careened around a corner, and the blur of bitumen came closer to my face than I ever expected, I had the out-of-body-like experience of wondering how misaligned gravity currently was with the bike. When a bike is standing upright, gravity pushes it directly into the ground and everyone is happy. Now there was some angle between the vertical orientation of the bike and the direction of gravity. 'I swear there is less than 45° between the bike and road,' I thought.

As a bike leans over, the direction of gravity becomes more misaligned with the body of the bike and that causes a turning force which, unfortunately, makes the bike tip even more. Thankfully the act of cornering brings some extra forces to the party which keep the bike balanced, but that involves a lot more physics. At the time, all I could think about was the angle.

I knew that gravity is always at right angles to the ground, so the bike–ground–gravity triangle must be a right-angle triangle, and thus if the bike–ground angle was under 45°, the bike–gravity angle must be over 45° to keep the total of all the angles 180°. See, not a great use of a triangle law, but a use all the same.

0.996195

In case you were wondering what the angle actually was, I got you covered. Looking up the bikes online, it seems the bike–ground angle can get as small as 30°. But I wanted to know my own personal angle, and therefore find out just how much my extra mass had limited the angle the rider sat in front of me could lean the bike into.

The lean angle is more complicated because of my tucking my arms in and moving around on the bike, but between the data from the phone and analysing the race footage I can confirm that the bike did lean just over 45° from vertical. Which means I can officially name my new motorbike gang Hell's Angles.

Heron's Formula

We started this chapter with one equation for the area of a triangle so I figure we can sandwich it by ending with another. This formula, named after Heron of Alexandria who lived in the first century CE, has some pros and cons over the previous area equation. For me, the main con is that I think the formula is just plain silly. It makes so little sense it almost makes me angry. But it works!

The downside to the classic 'area = ½ × height × base' is that it requires knowing the height of the triangle. Which is inside the triangle. That may be fine for something small like a sandwich, but when it comes to massive and solid objects, access to the middle of the triangle could be a problem. If you want to calculate the floor area of a triangular granary it would be preferable not to have to climb inside.

And, even if you can get inside a triangle, working out where the height needs to go is not exactly easy. It involves finding a line perpendicular to one of the sides which aligns perfectly with the opposite corner. Wouldn't it be great to have a second way to calculate the area of a triangle which only involves measuring the outside three sides, a, b and c? No problem. Heron's Formula mashes those three side-lengths together and, somehow, out drops the area of the triangle.

Start with the simple calculation of adding them all together: a + b + c. Then mix it up by adding two and subtracting the third, which gives you these three extra calculations: a + b − c, a + c − b and b + c − a. What if you now multiplied those four values together? If you do that and take the square root of their combined product, that is exactly four times the total area of the triangle. You might need to take a moment to sit down upon hearing that for the first time. I know I did.

$$\text{Area} = \sqrt{(a + b + c) \times (a + b − c) \times (a + c − b) \times (b + c − a)} \div 4$$

That's Heron's Formula. It winds me up, not because it doesn't work or we can't prove that it will always give the area (if you take the original area equation and hit it with Pythagoras's Theorem for long enough you can algebra your way to Heron's Formula). It bugs me because there is no obvious reason or neat logic for why it works. It's an opaque formula, and I feel like you just chuck in the side-lengths, turn a series of arbitrary mathematical handles and out pops the area. It works, but it's not satisfying.

I appreciate that for some people this is more wondrous than wrath-inducing. Which is completely valid. It's a kind of triangle party-trick, with a conjurer putting three side values into a calculator and pulling out an area. But mathematicians are rarely happy with just knowing something works. They want the beauty of understanding the logic behind the scenes.

0.998630

This is why I love triangles. A triangle is a surprisingly complex shape for a mere three sides. Somehow the humble triangle produces an incredible range of rules and properties, from the beautifully simplistic to whatever Heron's Formula is. And, somehow, they are always useful: if you need to work out the area of a triangular field or whatnot, don't worry about trying to align a height across the middle of it, just measure the sides and use Heron's stupid Formula.

It doesn't get better with more complex shapes either. Triangles are in the sweet spot of having enough sides to be a physical shape, while still having enough limitations that we can say generalized and meaningful things about them. Instinctively, I would expect shapes with more sides and angles to have such a plethora of rules and consequences. But in reality shapes with more sides get less exciting because there are so many ways they can be formed that it becomes almost meaningless to try and herd them all up into unifying laws.

So hurray for triangles and their many laws and patterns, of which I have explored a mere six here. Once we are armed with triangle logic, if complicated situations can be turned into a matter of triangles, then suddenly we have the entire might of the Laws of Triangles brought to bear.

Four

MESHING ABOUT

What happens when we deploy thousands, if not millions, of triangles all on the same problem? When we had a mere one or two triangles we were able to solve the problem of where a balloon is. Imagine what we can do with orders of magnitude more triangles at our disposal!

In mathematics we call a bunch of contiguous triangles (or indeed, any shapes) a 'mesh'. It is probably more akin to what regular humans would call a net – like a fishing net – but maths has already used 'net' for when you unfold a shape. And the word 'lattice' is reserved for when the shapes have much more order and regularity, which is a kind of mesh, but we want to have more freedom to mix and match all sorts of triangles into a flowing, flexible fabric. So we're going to be casting a fishing mesh instead.

Anything can be triangles. As in, you can take any surface of any shape and just absolutely cover it in triangles. Those triangles form a triangle-mesh approximation of the surface; the surface is now triangles. The size and number of those triangles will determine how close the new triangulated surface matches the original, and that is merely a matter of effort, will and budget. A rectangular mesh cannot match this level of versatility, regardless of how much budget might get thrown at it. Some things are mathematically impossible in a way that cannot be fixed.

We've already seen how architects are obsessed with rectangular windows, but that extends well beyond traditional cuboid buildings. Many modern buildings are far from flat, with all sorts of interesting building-surfaces now landing in cities around the world. Putting aside matters of structural stability, triangles and rectangles are equally good at forming a flat surface. But the moment you introduce any curvature, triangles once again take the lead. An architect is never happier than when they can 'panelize' a surface into rectangular glass panels, but there is a practical problem and a mathematical problem.

Practically, four corners do not necessarily form a flat shape. If all four points are not aligned in the same plane, then they cannot be joined by a flat sheet of glass. In contrast, any three points are by definition always in the same plane: triangles are always flat. A triangle mesh can always be filled in with glass, whereas a rectangle mesh probably cannot.

Mathematically, it is impossible to make almost all surfaces out of flat rectangles. There are only two types of surface which can be panelized. Firstly, donut surfaces can be made out of rectangles. Sometimes, when an architect slides over their sketch, the engineers can find a section of a torus which is close enough to what the architect wants, and do a quick Indiana-Jones-style swap-out for the toroidal chunk and hope no one notices. But the architects often do notice, and release a spherical boulder to chase the engineers down a tunnel.

Or there are surfaces that can look complicated but a keen eye will spot that the profile shape never changes. These are generated mathematically by taking two lines and combining them in a kind of 'curve multiplication'. Known as 'translation surfaces', they can work nicely in limited situations, but for a bigger construction do not allow much variety across the surface. Once the curvature has been decided upon, it needs to stay the same across the whole structure. But those are the only two options an engineer has if an architect insists on rectangles: donuts and translation surfaces.

A triangle mesh can do anything. Got yourself any surface? Bam: triangles. It's already done. Triangulating a surface only becomes a problem if there is a restriction on how many different triangles are allowed. This brings us back to the age-old tug of war between architects wanting shapes which are easy to manufacture and engineers wanting shapes which are easy to join together and form a strong building.

War of the Welds

In the early 2000s an architectural firm was designing a new high-rise hotel for Barcelona and they decided they wanted a UFO-shaped bar on the roof. The vision was a structure made from a lower concrete dish with a glass dome over the top, looking like a classic sci-fi UFO had just landed on top of the building, over 100 metres above the streets below. The concrete half was easy: you can pour concrete into almost any shape, reinforcement willing. But a glass dome is a bit harder.

The challenge of converting the dome from the architects' vision into physical bits of glass that will all fit together went to my engineering buddy Paul Shepherd. The designers had conceded that rectangles wouldn't do the job, but manufacturing triangle planes of glass is no small task (a real pane in the glass), so the builders stipulated that Paul should try and usc as few different triangles as possible. This way only a few different types would need to be made, and only a few different types of spares would need to be kept in reserve in case a pane needs replacing. If every triangle was different the spares would be a complete second copy of the dome!

I decided to see how I would solve this problem. I searched around for the 'most spherical' shape made out of copies of just one type of triangle and found the disdyakis triacontahedron (which means a shape with 120 faces made of thirty groups of four). You cannot get more identical triangles into a ball-like arrangement. A UFO has a clam-like final shape so we don't want half a disdyakis triacontahedron, just the top third or so.

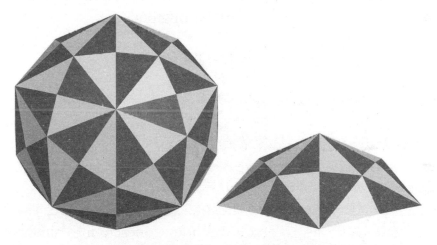

Top o' the disdyakis triacontahedron to you.

There are problems with this. Despite my spherical intentions, it looks pretty pointy and has too much of a pyramid vibe instead of a retro-futuristic look. And the triangles are big. Way too big: the distance from the bottom to the top is a mere two triangles. To make the bar big enough to have standing room, those triangle panels would need to be several metres long. That's bad in terms of the forces they will be under, and bad in terms of making the pieces of glass. Trying to make the UFO out of only one type of triangle ends up being too much of a reach.

Don't worry about the disdyakis triacontahedron, though, it still gets an application deserving of its record number of identical triangle faces: it's the world's biggest dice. 'Biggest' meaning largest number of faces. Triangles are the simplest shape, and the disdyakis triacontahedron is the maximum number of identical triangles you can fit in a spherical arrangement. The mathematical folks over at The Dice Lab have used the 'most triangular sphere' to make a 120-sided dice, for all your randomizing needs.

For when you cannot choose between 120 different things.

Paul knew he was going to need to use more than one type of triangle, so he started with an icosahedron, which is made of a mere 20 equilateral triangles. He then divided each of the faces into smaller triangles which could be 'lifted up' mathematically to make the shape much more spherical. This gives a far smoother dome but the process of inflating the mesh causes the edges of the triangles to stretch by different amounts, resulting in more than one type of triangle. The section Paul selected to make the UFO contained 105 triangles of six different types. For something so spherical, that's as good as you can get.

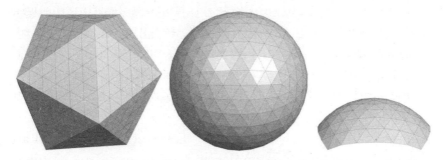

Covering an icosahedron with lots of triangles and then inflating it like a balloon.

But the designers were not happy with that. They were expecting all of the triangles to be the same! Paul then had to spend a week of his life not doing engineering, but instead coming up with ways to convince them that it was impossible to only use a single type of triangle. Eventually everyone decided they were happy with six types of triangle, and Paul got to work on the final design. Until there was one final curveball in the curved ball. They didn't like the edge.

Paul had envisioned the edge of the dome following the straight edges of the triangles, because if you 'cut it off' anywhere else you get fragments of triangles. His plan meant that the seam between the upper glass and lower concrete would not be completely flat: there would be small sections to fill in. The architects said, 'UFOs don't look like that.' Which is easy to laugh at, since it's serious architects arguing over the objective qualities of a science-fiction concept. But I get it. We all know what a classic UFO should look like.

Paul had to design another six glass shapes, some of them only sections of triangles, to fill in the gaps. And I know I make a lot of the architects-versus-engineers narrative because I find it funny, but the reality is that the back-and-forth between opposing forces tends to result in a satisfying blend of structural feasibility and overall aesthetics. I love modern architecture, so I'm very happy with this ongoing struggle between the two camps. I don't want all buildings to be boring cuboids, nor do I want fancy buildings which fall down.

I like to think of the Barcelona UFO as a mascot of the equilibrium which can be reached between beauty and structural integrity. Paul was so proud of the end result that he chose a meal in the UFO as the moment to propose to his now-fiancée. And you'll never guess what she does for a living. She's an architect.

0.996195

All Things are Triangle

During the discussion of how many different triangles the UFO would require, at one point the architects brought up the British Museum roof as an example for what can be achieved using only one, repeated triangle. Covering the largest indoor public square in Europe, this glass mesh is truly a testament to the power of triangles. But it represents the opposite of the architect's point. Disregarding the benefits of using as many identical triangles as possible, every single one of the 3,212 triangles in the British Museum roof is totally unique (with the caveat that the whole structure is symmetric, so each triangle has an opposite, reflected twin).

The roof designers picked a continuous surface that they wanted to make out of glass, and left it to the engineers to triangulize it. This was done using a computer to place neatly arranged triangles on the floor of the courtyard and then 'project them up', which means each corner floated directly up like a liberated helium balloon until it bumped into the ceiling. The code then allowed the triangles' corners to digitally relax and jiggle around to even out some of the discontinuities; after being updated 5,000 times they had all slid into a nice, even distribution. This process, known as 'dynamic relaxation', was also used to make sure the largest triangle did not exceed the largest possible glass size.

It was this step of letting the vertices jiggle about until they came to rest in the optimal positions that caused every triangle to be unique (excluding the side-to-side mirror symmetry). This must have been a deliberate engineering decision, as it would have been possible to constrain the number of shapes somewhat better than having 3,212 different ones. I'm glad they made that decision, as I think the freedom to have the surface look as fluid as possible has paid off with a spectacular piece of architecture.

But while there are many things this roof is an example of – a harmonious melding of engineering and design, an impressive new public space in a city as busy as London, something in the British Museum actually from the UK – there is one thing it isn't: a good example of what can be done with just one triangle. If anything, it's the perfect example of how complex a surface can be if there are no limits on how many different triangles are permissible.

I decided to put my maths where my mouth is and triangulate my face. The TrueDepth camera on the front of an iPhone projects tens of thousands of dots to scan 3D objects in front of it. These dots, which are designed to unlock your phone by detecting the shape of your face, are invisible to the human eye because they use infrared light. It is possible to export the 3D coordinates of all of the points produced when scanning an object, which means I can present to you: 3D me.

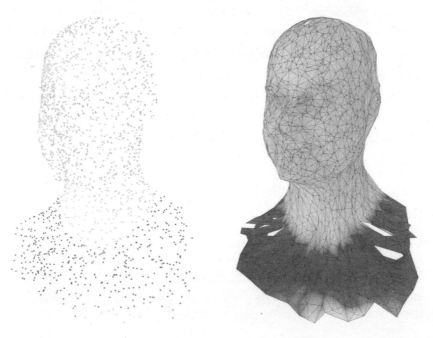

On the left is a collection of points in 3D space. On the right they have been joined up into triangles, recreating an accurate-but-terrifying version of my head. For any collection of points there are multiple ways they can be combined into triangles, and so algorithms have been developed to do this in 'good' ways. Delaunay triangulation is one such method, which specifically avoids small angles. You could say it 'maximizes the minimum angle' if you like confusing sentences. Keeping the smallest angles in a mesh as large as possible is handy because it reduces the number of 'sliver triangles'. These are really skinny triangles that reach right across the mesh in a way which makes working with it difficult.

I decided to go low-res and reduce my face to about 1,000 points, so the result would look properly triangley. Apart from the loss of my mouth and eyes, I still think it's not a bad approximation of my head (the approximation of the top of my head as a smooth sphere was not the fault of the triangles but rather nature).

Apart from this collection of triangles, which is one 3D-printer away from being the worst Halloween mask ever, you might be wondering what applications there are for turning a bunch of points into a virtual triangle mesh. Well, have I got three examples for you:

3D Printing

3D printing is all about triangles. In short: 3D printing involves taking a digital 3D model and sending it to a 3D printer, which will then start to build the shape up, layer by layer, until it gets out of whack and starts making an absolute mess before you notice what is happening, swear at it, restart the printing process and sit there watching it for hours this time to make sure nothing goes wrong. Or so I've heard.

The most common file format for 3D printing is an STL file and it seems no one is sure what STL actually stands for. Some say it is short for 'stereolithography', which means printing up in 3D layers, but I am much more partial to the theory that it stands for 'Standard Triangle Language'. Under the hood, an STL file is just a long list of triangles. You can actually open an STL file as if it were a text file and inside you will just see triangle after triangle. Each triangle is recorded as a series of three coordinates for the three corners, and then for convenience the 'normal vector' is also included (this is a vector which indicates which is the 'outside' and 'inside' of each triangle).

I contacted my friend Laura Taalman who is a Professor of Mathematics at James Madison University and a maths 3D-printing expert. She offered to send me one of her favourite STL files: a 3D model of a cube. A cube is very much not triangular, but I had a look inside the file and found a list of twelve triangles, which in pairs make up the six faces of a cube. It's such a small file I can actually include all of it in this book. Because the two triangles on each square face are perfectly flush, when I opened it in a 3D-viewing program it just looked like a perfectly normal cube.

```
solid OPENSCAN_model
      facet normal -0 0 1
            outer loop
                  vertex 0 1 1
                  vertex 1 0 1
                  vertex 1 1 1
            endloop
      endfacet
      facet normal 0 0 1
            outer loop
```

```
            vertex 1 0 1
            vertex 0 1 1
            vertex 0 0 1
        endloop
endfacet
facet normal 0 0 -1
    outer loop
            vertex 0 0 0
            vertex 1 1 0
            vertex 1 0 0
        endloop
endfacet
facet normal -0 0 -1
    outer loop
            vertex 1 1 0
            vertex 0 0 0
            vertex 0 1 0
        endloop
endfacet
facet normal 0 -1 0
    outer loop
            vertex 0 0 0
            vertex 1 0 1
            vertex 0 0 1
        endloop
endfacet
facet normal 0 -1 -0
    outer loop
            vertex 1 0 1
            vertex 0 0 0
            vertex 1 0 0
        endloop
endfacet
```

```
facet normal 1 -0 0
    outer loop
        vertex 1 0 1
        vertex 1 1 0
        vertex 1 1 1
    endloop
endfacet
facet normal 1 0 0
    outer loop
        vertex 1 1 0
        vertex 1 0 1
        vertex 1 0 0
    endloop
endfacet
facet normal 0 1 -0
    outer loop
        vertex 1 1 0
        vertex 0 1 1
        vertex 1 1 1
    endloop
endfacet
facet normal 0 1 0
    outer loop
        vertex 0 1 1
        vertex 1 1 0
        vertex 0 1 0
    endloop
endfacet
facet normal -1 0 0
    outer loop
        vertex 0 0 0
        vertex 0 1 1
        vertex 0 1 0
```

```
            endloop
        endfacet
        facet normal -1 -0 0
            outer loop
                vertex 0 1 1
                vertex 0 0 0
                vertex 0 0 1
            endloop
        endfacet
endsolid OpenSCAD Model
```

Yep, that is a cube.

To print something more complicated than a cube the 3D models get more involved. But they are still lists of triangles – just a lot more of them. For my previous book, *Things to Make and Do in the Fourth Dimension*, Laura very kindly printed three interlocked rings for me (known as Borromean Rings, they have the property that if you remove any one ring, the other two come apart as well). She opened up the STL file for me and counted a total of 19,656 triangles.

There is of course some clever software which can take an STL file and turn it into layers for a specific 3D printer to be able to actually make it, with all sorts of clever infill and supports. But the model itself is always just a list of triangles. I checked, and Laura has a model online for printing the frame of a disdyakis triacontahedron. Of course she does. The shape itself consists of 120 triangles, but to 3D-print all the edges will run you 115,200 triangles. And there is an option to get just the middle band of the shape 3D printed professionally in gold (-plated bronze), for the nerdiest bracelet you'll ever wear.

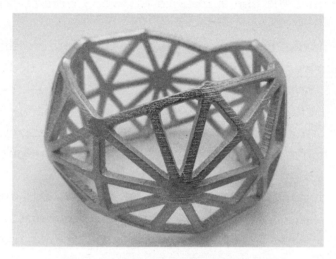

Triangles made of triangles made of gold.

Visual Effects

This will be easy, I thought. I have another friend, Eugénie von Tunzelmann, who works in visual effects. Same as Laura, I'll email her and ask nicely for some good examples of triangle meshes being used in VFX, maybe reference a film or two everyone has heard of, and job done for this section. But Eugénie's reply was full of, while not entirely 'bad news', definitely novel news.

We never want to use triangle meshes in VFX, ever.

I hadn't been this shocked since Heron's Formula. I had this whole section of the book planned out! In my defence the video game industry does use all triangle meshes in their computer-generated imagery (CGI). Triangles, with their mere three sides, are quick to compute and video games are all about how many frames-per-second of CGI can be crunched out. But with a film there is the luxury of time: all of the visuals get rendered once and then do not need to change once the film is released (unless George Lucas gets his way).

It turns out that when CGI is being made for the film and TV industry all of the meshes are based on four-sided quadrilaterals, not triangles. These 'quad meshes' take longer to render but they produce better visual results for a number of reasons. I'm not too stuck in my triangular ways to acknowledge when a different type of mesh is more fit for purpose, so here are some useful things about quad meshes.

The big one is that, with their pairs of opposite sides, quadrilaterals stack nicely. It's easy to form a long chain of quadrilaterals with neatly aligned edges in a way that triangles struggle to replicate. These chains of shapes in a mesh form what are called 'edge loops' and, for a good realistic render, edge loops will ideally be aligned along the edges, boundaries and folds in a surface to keep them looking sharp. Quad meshes are better at being neatly aligned with the contours and edges of physical objects, which makes for a more realistic approximation of their surfaces.

Quad meshes also respond better to a process called 'surface subdivisions', which is the 'zoom and enhance' of the CGI world. There are several algorithms available with silly names like Loop, Catmull-Clark, Modified Butterfly and

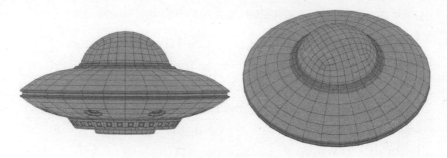

Starship Loopers: a quad grid of a CGI UFO with neatly arranged edge loops.

Kobbelt but they all have the same goal: split the mesh into even smaller shapes to increase the resolution of the final render. Quad meshes respond better to this process and give much smoother results.

Finally there is tradition. Early 3D surfaces in VFX were made exactly like the translation surface engineers use when they have to make a surface out of rectangles. This process of multiplying two curves always gives a quad mesh, and was used a lot when it was not computationally feasible to just convert any arbitrary surface into a mesh. This is now baked in, and the industry continues to prefer quads and has better developed tools for working with them. The more charitable angle is that quad grids are easy enough to produce and manipulate that they allowed the world of CGI to get started in the first place.

But there is a price to pay for being the only quad-mesh industry in town. Eugénie explained that because 3D modelling and computer games all use triangle meshes it is common for a client to have a pre-existing 3D model, or for an ideal stock-model to be found in an asset library, but it is all completely triangles. While there are of course techniques out there to 're-topologize', and turn a triangle mesh into a quad mesh, she explained that this is not so easy, and is actually a bit of an art that requires manual adjusting of the mesh. VFX people do not want any old quad mesh; they want a good one with sensible edge loops and distribution of the quads and that can take some effort. (Only in extreme cases, like a particle-based water simulation, would a triangle mesh be used, and then only if the triangles in the mesh were so small as to be almost invisible.)

Then came the final bombshell: VFX artists do not care if their quads are planar – flat, in other words – or not. Engineers very much care because they need to fit a piece of glass, or some other really-prefers-to-be-flat object, into their meshes. But in the virtual world it doesn't matter.

> we allow 'non-planar quads', which we still consider to be quads from a data point of view, but mathematically . . . well that's just two triangles, isn't it?

YES IT IS EUGÉNIE. After all this talk of quad meshes, Eugénie let me in on the great irony that they were actually just triangle meshes all along. Albeit a special type of triangle mesh where all the triangles are buddied up with a partner to form a quadrilateral, which is certainly not possible in just any, arbitrary triangle mesh. So, I get it, the quadness is the definition factor of the mesh. But when the grid is actually processed and rendered the computer treats every quadrilateral as two separate triangles.

I wanted to know how many triangles but, for not the first time ever, a friend told me that I cannot use their work as an example in my book because they are under a nondisclosure agreement. Despite her award-winning work on the *Jurassic World* franchise, Eugénie could not tell me how many triangles the T-rex uses. The 'Triangulosaurus rex', if you will. But she did disclose that, generally speaking in a modern film, something like a human character might run to about half a million shapes in their mesh. Which does explain why my 1,000-triangle selfie mesh looked so uncanny.

Engineering

Engineers also triangulate surfaces, not because they want to build the structure out of triangles, but rather to double-check it's not going to fall over.

'Finite element analysis' is the process of splitting a structure or object up into composite parts which can be individually analysed. Humans have had a pretty good understanding of forces for a few centuries, but engineers have been limited in how they can be applied to whole structures because of a lack of computing power. Thankfully, computing power is something computers are increasingly good at providing. It is no coincidence that the rise of non-cuboid architecture has coincided with the era of electronic computers.

In a perfect world some kind of supercomputer could calculate all of the forces as they vary across an entire structure (maybe going atom by atom), but for now we need to split a building into small-enough-is-good-enough chunks. Think of these finite elements as like the 'pixels' of a building, engine part, wing or whatever object we need to calculate the forces of.

Here comes a familiar kicker: not all finite element analysis is done with a triangle mesh. Sometimes a building or beam is very rectangular in nature, and so a quadrilateral mesh fits better. Very similar to the edge-loop problem in VFX, it can be easier for a quad mesh to align with the orthogonal edges which are so popular in buildings. So chalk another one up for the Quad Squad. These are proper quadrilaterals as well, not the cheeky two triangles of VFX.

But this is only the case for certain structures. If an engineer is analysing a part of a machine or building which has any kind of complexity to its shape, it's a triangle mesh all the way.

I contacted Paul again and asked if he had done the finite analysis on any structures he could share with me (which were not under an NDA!) and he sent me the files for some hexagonal concrete shells that form a kind of shade umbrella at a train station in Stuttgart, Germany. Because they were symmetric along one axis, he only had to design half of it and then flip it over to get the other half. And because it was a flowing, organic shape, everything was done with triangles.

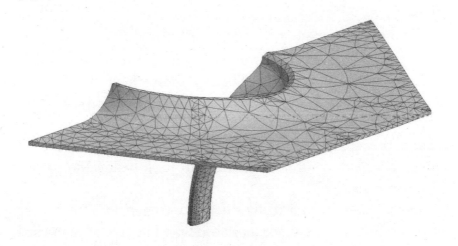

More than the standard six triangles required to make a hexagon.

What you cannot see from the images is that, actually, everything in that hexagonal concrete structure has been split into tetrahedrons, 3D triangles! Not only was the surface mesh made of triangles, but the entire 3D mesh that filled the whole structure was triangles all the way down.

Noise and Easy

In 1997, random noise won an academy award. I'm using 'noise' in the mathematical sense to mean a random signal which could be audio or visual. The black-and-white static mess on a detuned antique TV is an example of noise. In the modern world, the weird colour patterns you see when a digital signal glitches can be considered noise. It is basically any kind of meaningless but busy signal.

Noise winning an academy award seems like a scathing review of modern motion pictures, but it is completely true in a technical sense. Professor of computer science Ken Perlin was working on the 1982 film *Tron* but was dissatisfied with the 'random' noise being used in some of the visual effects. We spend our lives swimming in an ocean of randomness, so much so that we often forget it is even there. Until we need to simulate reality from the ground up using computer graphics. Then suddenly everything looks really artificial until some realistic randomness is applied. The position of leaves in a tree, the texture of a worn footpath, the slight wobbles in a hand-drawn line, these all need an element of randomness to look believable to our human eyes.

Ken worked out a really nifty way to generate believable random noise to enhance digital effects. The Academy Award he won was not for any specific film, but rather for the concept behind his noise in general. He received an Academy Award for Technical Achievement (aka the ones which are not televised) for 'a technique used to produce natural appearing textures on computer generated surfaces for motion picture visual effects'.

I asked my VFX friend Eugénie how often she used Ken's random noise in her work. She said it is used so often it's hard to give any one example, it's so ubiquitous. She then rattled off a series of examples from the top of her head: 'to make a car's shiny paint a little rougher, so it doesn't look so perfectly new. Or we might use it to scatter trees onto a landscape, and again to scatter snow onto the mountaintops. We could use it to shape and colour rocks.' Modern film and TV special effects cannot exist without good random noise.

It may seem like the answer to this problem is just 'use random numbers'. And look, I'm a big fan of random numbers, and if this problem could be solved by rolling a 120-sided dice over and over I'd be there. But it can't.

When digital artist Seb Lee-Delisle was programming a laser installation, he needed a realistic way to generate strikes of lightning. For lightning to look real it couldn't be one straight line: it had to zig and zag back and forth a bit. Pure random numbers would make for a very disjointed bolt as it skipped from side to side. Natural randomness has a bit of smoothness to it, changing value in a random but continuous way. That takes some cleverness to recreate. Seb also wanted to loop the animation of lightning moving, so there would need to be a way to get a sequence of randomness to return perfectly to its starting position, a decidedly non-random thing to do.

Perlin's insight was to find a way to generate an entire landscape of randomness that people could pick and choose from according to their randomness requirements. Think of Ken as a farmer growing a huge field of all-natural randomness, and anyone who needs some can harvest exactly the path of randomness they want. In Seb's case, a circle of randomness that comes back to where it starts, and would therefore begin and end with the same value. You could imagine doing this on an actual landscape: going for a hike and keeping track of your exact elevation as you go through hills and valleys. This is what Perlin noise is, but with the advantage that the landscape is mathematically generated and endless.

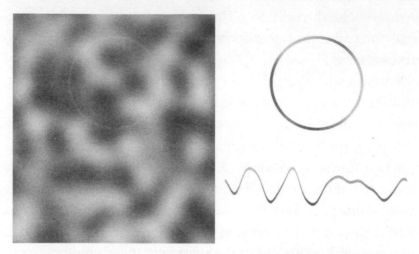

On the left is a field of Perlin noise and a circle of noise has been extracted from it. If we 'unwrap' that circle we can plot its values to give us a nice and natural random signal.

For those who want to meander through the details, imagine that as you walk this mathematical landscape you notice it has been completely divided into square fields by an endless grid of fences. You look at the four corners of the field you're currently in and notice they have all been labelled with a random value. To get your local randomness value you calculate the distance to all four corners, multiply each distance by the random value at the at corner and then use a clever calculation to combine the four results into a single final value (this involves vectors and a bunch of other fun stuff we don't need to worry about).

This was Perlin's genius idea. The value at any location depends on the four nearest grid corners and how far away they are. But they are scaled so that as you move away from one corner its influence gradually wanes. Once you hop a fence into the next field, the influence from the two distant corners in the field you're leaving reaches zero, and the two far corners in the new field start to weigh in. Going from field to field, the influence from the corners of the grid gradually increases and decreases as you move towards and away from them. This gives a very smoothly changing surface of randomness.

When I was recording a podcast about randomness with my musical friend Helen Arney, I wanted to talk about Perlin noise but was limited to audio only (which I appreciate is the opposite situation to writing a book). I chose a circular path through the Perlin fields of noise, and mapped the random values to a music scale to produce a random tune. Because Perlin noise changes smoothly, this gave a much more natural and flowing random tune compared to one where I picked the notes completely at random. I'm totally amusical, but if you're like Helen and understand the following symbols you'll also be able to experience the difference. Or take this book, run over to your nearest musical friend or family member and shout 'Sing these!'

True random tune:
A4, A3, A4, A3, F#4, E5, B4, D5, C#4, C#5, C#4, G#4, B3, G#4, A4, B4, F#3, F#4, G#3, E5, A4

Perlin random tune:
A4, D4, B3, A4, A4, F#4, B3, D4, E4, C#4, D4, G#4, F#4, A4, E4, B3, E4, D4, B3, D4, A4

This was a flexible system, as I could move the circle path around to get different tunes, or change its size to allow for

0.927184

more notes, or for the notes to be closer together or further apart. The only thing I couldn't do was generate a series of tunes which all differ slightly from the previous one and then return back to the original starting tune. Which is exactly the same problem that Seb had with his lightning lasers. This is a common problem in the world of organic noise, and one that can be solved with the third dimension!

With three dimensions of freedom, a circular path can itself be moved around in a circle. If that is hard to picture, imagine a donut floating in the air. A circle moving around a circular path will trace out the shape of a donut. But to make this work we're going to need some triangles. So far, Perlin noise has been suspiciously all about a square mesh, but we're going to need triangles if we go any higher.

Perlin noise can be extended to the third dimension but it requires a 3D cube lattice. Which makes for a lot of extra corners, and each extra corner introduces another complicated calculation. This is the advantage that triangles have over squares: they have fewer corners. Which holds for any dimension! The 3D version of a square is a cube with eight corners. The 3D equivalent of a triangle is a triangle-based pyramid (aka a tetrahedron) and that has a mere four corners.

While humans have no experience with the fourth spatial dimension, the maths is still pretty straightforward. I've drawn approximations below for fun, but you don't need to stress about trying to visualize what these hyper-dimensional shapes look like. For any dimension 'n' there is a square equivalent shape called an 'n-cube' and a triangle equivalent called an 'n-simplex'. The triangle shapes increase at the rate of one corner per dimension whereas the number of corners on a square doubles every time. The moral is, triangles stay manageable whereas squares explode.

Two dimensions

3 corners

4 corners

Three dimensions

4 corners

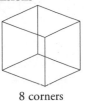

8 corners

Four dimensions

5 corners

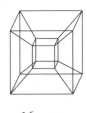

16 corners

Five dimensions

6 corners

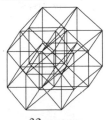

32 corners

After Ken Perlin came out with Perlin noise in 1983 he had a good think about it and, in 2001, unveiled the sequel: simplex noise. It had a few inconsequential upgrades compared to Perlin noise (and now shuffled the values at the corners instead of being pseudorandom), but the big change was moving from a square grid to a triangular grid. A mesh of 3D, 4D or even higher triangles was now available so anyone could access noise with any level of complexity they may require. I used the 3D version of simplex noise when I was making the random tunes for Helen.

A4, D4, B3, A4, A4, F#4, B3, D4, E4, C#4, D4, G#4, F#4, A4, E4, B3, E4, D4, B3, D4, A4

A4, D4, A3, G#4, C#5, G#4, D4, C#4, D4, D4, F#4, G#4, C#4, B3, E4, E4, B4, G#4, C#4, D4, A4

A4, D4, G#3, C#4, E4, E4, D4, C#4, A4, D4, E4, E4, F#4, A4, E4, B3, A4, A4, D4, E4, A4

0.913545

A4, D4, A3, C#4, A4, F#4, G#4, G#4, A4, B3, C#4, C#4, A3, G#4, A4, F#4, A4, G#4, D4, E4, A4

A4, D4, C#4, F#4, A4, C#4, G#4, E4, E4, G#4, B4, D4, F#4, F#4, A4, B4, F#4, G#4, E4, E4, A4

A4, D4, E4, F#4, G#4, A4, G#4, G#4, D5, G#4, A3, C#4, A4, G#4, E4, B3, A3, E4, E4, E4, A4

A4, E4, E4, A4, B3, E4, G#4, B3, C#4, F#4, F#4, D4, G#4, E4, C#4, F#4, A4, D4, E4, D4, A4

A4, E4, E4, F#4, B3, G#4, A4, E4, C#4, A4, A4, A4, C#5, E4, E4, A4, C#4, D4, E4, D4, A4

A4, E4, C#4, E4, D4, F#4, G#4, G#4, E4, F#4, E4, E4, C#4, B3, E4, C#4, D4, B4, E4, C#4, A4

A4, E4, C#4, G#4, F#4, C#4, E4, B4, A4, D4, D4, F#4, E4, C#4, D4, E4, E4, E4, C#4, C#4, A4

A4, D4, B3, A4, A4, F#4, B3, D4, E4, C#4, D4, G#4, F#4, A4, E4, B3, E4, D4, B3, D4, A4

This series of tunes all start and end on an A note because they were all circular paths through the noise. This circle then swung around the surface of a donut, which means each tune changes smoothly from the first one and eventually loops back to where it started without any repeated tunes along the way. It's very mathematically clever but that does not guarantee any musical quality. Tunes generated from a donut surface are not guaranteed to sound at all sweet.

My terrible random tunes aside: this is exactly what Seb

was doing with the lasers. As the watching public were wowed by the bright, interactive laser lightning, little did they know they were actually watching a circle swing around a donut plucked out of a 3D field of randomness. Likewise for all manner of digital effects where simplex noise is the unsung (and occasionally sung) hero. Because a mesh of triangles can more efficiently fill higher-dimension space with fewer vertices than squares, all sorts of complicated paths through fields of noise are within computational reach. A VFX artist can start with a 3D blob of randomness, move it around however they fancy and get back to where they started, potentially along multiple different paths that never cross. If that doesn't deserve an Academy Award I don't know what does.

Colourful Language

You would think the act of selecting a photo on your phone and printing it would be fairly straightforward. The phone just needs to transmit that image to the printer, which then does its printing thing. But the issue is that phones – indeed all digital cameras – speak a different colour language to printers. And there is a device argument over whose problem the translation should be. An argument which is solved with a triangle mesh.

When a digital camera takes a photo, it doesn't store that image as a list of pixels with colour names like 'red', 'blue' and 'puce'. It's a digital device, which means it considers representing colours as anything other than digits completely maroonic. Instead, it uses numbers. I've talked at length in previous books about how digital images store each pixel as a set of Red, Green and Blue values, but I never bother talking about what the red, green and blue actually are. And it changes from device to device.

If you take a photo on your iPhone it will be saved in

a colour encoding called DCI-P3. When you send it to your computer and look at it on a screen it has probably been converted to sRGB. When the whim to print it takes over, the photo will be converted once more, to CMYK, before it's sent to the printer. Each of these different ways of representing colours is called a 'colour space', and converting from one colour space to another – working out the equivalent colour encoding for each colour in the image – is a surprisingly complicated calculation.

The computer intermediary in the above hypothetical situation is the secret to success. It can listen to the iPhone talking in DCI-P3 and convert it over to CMYK so the printer can understand what it's saying. But printing a photo directly from a phone to a printer without a computer in the middle poses a problem: neither the phone nor the printer is designed to do difficult mathematical computations. They have pretty light-weight processors which would struggle to translate between colour spaces.

This problem first presented itself in the early 2000s and you might think that as phones get more powerful, the problem is solved. But we just keep taking higher resolution photos. Plus neither device wants to do it! The phone thinks that colour spaces are a printer problem, and printers think the phone should send it the final file ready for printing. All of that aside, we can all agree that if the tedium of preparing an image for printing can be done with less processor time, that capacity remains free for more important things, like taking even more selfies.

The solution was twofold. First, it was agreed that sRGB would be the communal middle language. It is a good standard version of RGB insomuch as everyone assumes that the 's' stands for standard. But I directly contacted the International Color Consortium, who are in charge of these

0.891007

things, and they confirmed the 's' has no official definition. So in my headcanon 'sRGB' stands for 'supercalibratinglogisticredgreenblueydocious' and you can't stop me.

The second step was to do all of the calculations in advance. Instead of having to do the computationally intensive conversion for every pixel, every time, someone like HP could just produce a huge table of all 1 billion DCI-P3 colours, and all a phone would need to do is look up the matching sRGB value for any given pixel. And, in 2002, HP employee Peter Hemingway released a research paper called 'n-Simplex Interpolation' which cracked the colour-conversion case wide open. Peter had found a way to pack the required conversion table into a fraction of the expected hard drive space, and just in time for the dawn of the mobile phone era.

Storing every possible colour in a massive table will take up loads of drive space on a phone. So, what if HP made a table for every second value, and if a phone needs an inbetween value it can just average the two adjacent values? Or only put every fourth value in the table, and the values between can be interpolated by the phone. This is a great plan because it means the table can be much smaller, and the interpolation calculations are far more simple than the original colour space conversions. There are two reasons those interpolation calculations can be efficient: hard work from HP and triangles.

We need a quick method that can take two known values and estimate any unknown inbetween value. Logically, the closer a point is to a known value, the more similar it should be. The key insight deployed by HP was that as an unknown point moves closer to one known value, the distance to the value on the other side increases. This can be used to do a 'weighted average' where each known value is multiplied by the distance on the other side. If we know values 'A' and 'B' and want to estimate the value 'x' in the middle, this is the weighted-average calculation:

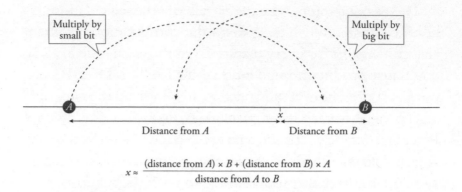

Multiply by small bit

Multiply by big bit

A

x

B

Distance from *A*

Distance from *B*

$$x \approx \frac{(\text{distance from } A) \times B + (\text{distance from } B) \times A}{\text{distance from } A \text{ to } B}$$

This all works nicely if you need to interpolate between values which sit in a neat, orderly line. But colours are made up of three values, like the red, green and blue values of RGB. This can be thought of as a 3D system of values, with R, G and B each getting their own axis. I find this a very pleasing way to visualize the RGB, and indeed any three-colour colour space. It is a fantastic coincidence that humans both have three colour receptors in our eyes and experience a three-dimension spatial reality.

Imagining R, G and B as spatial dimensions means that conceptually the whole table of RGB conversion values is just a giant 3D cube lattice. This can be a bit hard to picture, but it does reveal how our geometric tools can be used to tame a large conversion table. It brings order to what would otherwise be a long list of numbers with no rhyme or reason.

If you're prepared to stretch your ability to visualize 3D space, the same interpolation method from before still works in higher dimensions. For 2D, each of the corner values is weighted by the area of the opposite section, which is easier to picture than the 3D case we actually need. Interpolating the actual 3D point requires weighting by the volume of the opposite tetrahedron.

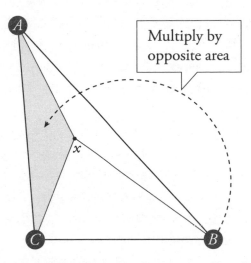

A, B and C are all multiplied by the opposite area and then the total is divided by the total area.

Don't worry, I'm talking about cubes but showing triangles deliberately. It's because Peter Hemingway at HP had the same insight as Ken Perlin: squares and cubes have too many corners. If a missing colour value can be interpolated from a tetrahedron instead of a cube, then only four input values need to be crunched instead of eight. The method he developed picks the closest four corners of the surrounding cube, and deals with them as a tetrahedron.

This process has to happen fast and with the minimum possible calculations. We need those in-between values with minimal processor load otherwise I'm going to have yet another reason to want to throw my printer out the window. Peter needed a way to quickly calculate the volume of a tetrahedron. Actually, Peter needed to be able to quickly calculate the area of a triangle, the volume of a tetrahedron or indeed the 'content size' of any simplex. He wanted this

0.866025

method to work for any data application in any number of dimensions. And what method did he use to find these volumes super fast?

I kid you not. HERON'S AREA FORMULA.

I'm sorry. This is a very emotional chapter for me. First Eugénie's rollercoaster-ride of reveals, and now Heron's Formula has slammed into the room with a practical application.

For 2D data extrapolation this method uses Heron's Formula exactly as we previously saw it, and there is a 3D version of Heron's Formula which can calculate the volume of a tetrahedron using only the six side-lengths. And it does not involve calculating any angles, just doing simple operations on the side-lengths, making it really quick to compute.

Dear reader, I need you to believe me that I had already written the chapter about triangle laws where I called Heron's Formula stupid (because it is) when I read the official HP documentation for this technique, and came across the sentence 'The area of a triangle is given by Heron's formula.' I honestly just pushed back my chair, stood up and silently left the room to go for a walk outside.

The next time you print from your phone, know that both it and your printer are searching through a tetrahedron mesh and applying Heron's now-no-longer-as-stupid Formula.

Ballooning Applications

I think we should end this chapter with some uplifting news. Peter Hemingway's colour-conversion interpolation method was published with the title 'n-Simplex Interpolation' to make it very clear that this was a technique that could be applied to all manner of situations. Sure enough, this triangle mesh consisting of potentially millions of triangles is being used to . . . work out where a balloon is.

A researcher from RMIT University in Melbourne, Australia got in touch with me because they use n-Simplex Interpolation to track high-altitude balloons (used as a kind of pseudo-satellite). They need to factor in things like wind vectors, temperatures, pressures and densities to keep track of where a balloon has gone after launch. Of course, they do not know all of those values for every possible location in the sky, so they interpolate them using, of all things, a four-dimensional tetrahedron mesh.

They explained to me that, 'If done right, ascent rates can be targeted to within centimetres per second, and positions to within several hundred meters after hours (and a few hundred kilometers) of flight.' So even though we now have orders of magnitude more triangles and they exist in a virtual 4D mesh, they are still being used to work out where a balloon is. And somehow Heron's Formula has come along for the ride.

Five
WELL FIT

I have already mentioned a shadowy secret society of trianglists (to explain the omnipresent Pythagorean Theorem) but if there is a real conspiracy going on it would absolutely be the Hexagon Conspiracy. I stand by my choice of the triangle as the face of geometry, but the hexagon is definitely a key player. Everywhere you look, there are hexagons where there absolutely should not be. The rock formations of the Giant's Causeway in Northern Ireland are made of regular hexagons. The planet Saturn is wearing a hexagon as a hat. Snowflakes are tiny, magical hexagons. And bees! Bees are all about the hexagons.

I find it bizarre that something as perfect and precise as a regular hexagon appears seemingly spontaneously throughout our universe. Certainly more than we see regular triangles appearing of their own accord. Humans have also been using hexagons since the dawn of civilization; ancient Roman tiles have been found which are regular hexagons. Our cutting-edge technology also continues to contain hexagons: graphene is a hexagonal grid, and the JW Space Telescope has a primary mirror made of eighteen hexagons. Can this all be a coincidence?

Hexagons in spaaaace!

But you look into the geometry of it all and it turns out that bees, Romans and NASA engineers are using hexagons for exactly the same reason. In NASA's words, hexagons are the hexagreatest because they have 'high filling factor'. Which is to say they fit together real nicely.

When the JW Space Telescope was launched in 2021 it became the biggest telescope in space. So big it had to be split into smaller mirrors for easy transport off the Earth. Once in space, those mirrors had to clip back together, and hexagons are the best way to reduce the number of edges. When light hits the hard edge of a mirror it diffracts and spreads out, the opposite of the focusing in you want a telescope to do. Hexagons minimize this, but not completely without making their presence felt: if you look at the images the JWST produces, the stars all have six spikes coming out of them. Those six spikes are the diffraction patterns from the six edges of the hexagonal mirrors (fun fact: the Hubble Space Telescope uses one circular primary mirror, but produces four-spike stars because of the four arms holding the secondary mirror in place).

Bees were faced with the problem of how to build honeycomb with maximum capacity for minimum wax. Ancient

Hexagons filling spaaaace!

The six-pointed stars are the result of the hexagonalness of the mirrors.

0.819152

tile makers wanted tiles that didn't require too much grout. It's all the same problem: what shape fits together neatly with minimal seams? The same problem that always has the same solution: hexagons. Not only do hexagons fit together with no gaps – a property they share with squares, triangles and countless other shapes – but they are absolutely the most efficient shape possible.

I previously called out non-right-angle triangles as just two right-angle triangles in a trench coat, and I actually view a lot of shapes as mere collections of triangles. A hexagon can be seen as six equilateral triangles who are all just really good friends. Triangles are the foundation of all other geometry. But obsessing over just triangles can mean we miss the wood for the trees. Other shapes may be built of triangles, but once assembled they take on their own new, and sometimes unique, properties.

This is why astronauts and bees use hexagons instead of just equilateral triangles. A hexagon gives you an extra 50 per cent more area for the same perimeter. Shapes are more than the sum of their triangular parts. So it's worth taking a moment to explore some other shapes that triangles make possible and see how they all fit together.

Taking Care of Bees-ness

It is an oft-repeated claim that bees are good at maths. For centuries, people have even tried to use geometric structures built by bees as proof of some kind of deity who must have schooled them in angles. But I'm going to say it right now: bees are not doing geometry. They're not getting out little protractors and measuring angles. As adorable as that would be. They are making hexagons by accident. The same way

Saturn and the lava which formed the Giant's Causeway produced hexagons.

If bees are not doing maths, then what are they doing? To find out I asked a bee person. Vincent Gallo is a retired software developer who did a PhD in bees at Queen Mary University of London. To investigate the logic bees are following to build their honeycombs, he does two things: he watches as bees build honeycomb normally, and he gives the bees strange starting conditions (that is to say, some preformed wax which is shaped unusually) to see what the bees will do.

Here you can see two of Vincent's photos of the same bit of honeycomb. On the left it is still under construction and some of the openings look weirdly circular. Left to their own devices bees actually build circular cells, not hexagonal ones. That's right: a single bee cell would be a cylindrical tube of wax. But bees don't build just one, they build many cells all next to each other. It's the interaction of adjacent cells which causes hexagons to form.

Wax is a flexible substance and bees take advantage of that by constantly pushing it around as they build, as well as scraping it back up and reforming it. While a bee building a cell will push all of the walls out to make a cylinder, a bee in the cell next door is pushing back. And it is that back-and-forth pushing of the malleable wax which causes the cells to eventually settle into hexagonal shapes. Hexagons are a shape which can naturally result like this, which is why they pop up so often in nature.

This back-and-forth pushing of the wax allows adjacent cells to settle into a kind of equilibrium where the space is split evenly. Bees are not building hexagons and they have no idea what a 120° angle is. They have found a physical system

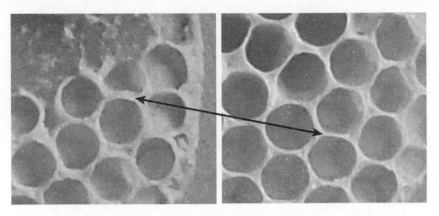

Same wall, before and after bees complete the other cells above it.

which divides angles in half. When Vincent gives the bees a strange starting angle they will build a new wax wall and then push it around until it reaches an equilibrium exactly in the middle. You can even see this on the edges of honeycomb where bees are building off a flat wall: perfect 90° right angles all the way along. Bee plus for effort.

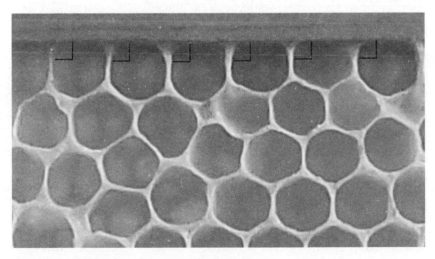

Right on.

Tile and the Whole World Tiles with You

The concept of making shapes fit together without any gaps is called 'tiling' in mathematics, named after the actual tiles humans have been using to decorate buildings for at least 3,000 years. Going back further, the ancient Sumerians were using repeating patterns to decorate buildings 5,000 years ago and there are mammoth tusks over 10,000 years old which humans etched repeating hexagon patterns into. Like bees, humans have an innate wish to arrange things in neat repeating patterns, but unlike bees we can choose to use all sorts of interesting shapes.

It makes me sad that modern tiles and paving patterns seem to be so obsessed with rectangles. They work fine but we can do so much better. When we were having a courtyard behind our house resurfaced, I decided to pave the way with new and exciting paving patterns. (My wife Lucie was happy as long as I did all of the negotiations with the builders). I did a quick stocktake of all the shapes which tile in a nice regular pattern (often called a 'tessellation') and found I was spoiled for choice.

A flat shape with straight edges is a polygon, and I had decided to limit myself to only polygons as I didn't want to complicate things by allowing curved edges. This still left ample options. Any kind of triangle can be arranged to completely cover a surface. Not just the equilateral triangle, any triangle whatsoever will do the trick. Likewise, any four-sided quadrilateral can be repeated over and over to form a tiling pattern. Including the ones with 'concave', sticking-in edges. If you receive a shipment of three- or four-sided tiles, it does not matter what shape they are, as long as they are all identical they'll definitely cover a bathroom wall.

I figured that having concave polygons would make

cutting the shapes out of stone needlessly complicated, so I committed to finding 'convex' polygons only. In one swoop this removed all polygons with seven or more sides. Not a single convex polygon, from heptagons up, can tile a surface. Which left me with pentagons and hexagons.

There are only three families of convex hexagons that will tile nicely, and they are shown in the following diagram. Each has their own set of constraints: sides which have to be the same length and angles which need to give a specific sum. The third case is subtly different because you need to flip some hexagons over to get a mirror image. Interestingly, the regular hexagon (with all sides and angles the same) fulfils the criteria of all three categories. I think of the regular hexagon as the canonical tiling hexagon, and these are three different ways it can be distorted while still being able to tile.

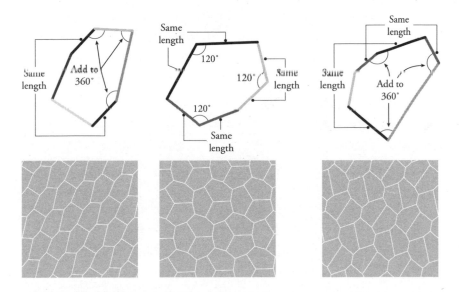

I was chatting to one of my Patreon supporters (an elite group of maths lovers who help fund my YouTube videos) named Timon, and they told me how they and their partner were both so mathematically inclined they decided to have their wedding cake decorated in hexagons. As the ultimate tiling option, the hexagon is the perfect shape to both represent a marriage where everything fits together nicely and cover a cake with no gaps. They even sent an example render image to the baker showing how the cake would look with regular hexagons neatly covering its cylindrical surfaces. What they saw on their wedding day, however, was very different.

I would not be able to hold my peace.

In a replay of the octagon biscuits labelled as hexagons, the person making the cake heard 'hexagons' but made octagons. What is it with bakers and mixing up hexagons and octagons? As already established, octagons are unable to tile a surface no matter how hard you try, so the baker just tried to jumble them all together. Thankfully, Timon and partner have a sense of humour and appreciated their wedding cake all the same, enjoying the thought of an angry baker cursing the octagons for mysteriously refusing to fit together like the hexagons in the render they were looking at. At last, some consequences for not paying attention to the names of shapes!

In my paving-tile hunt I left pentagons until last because it's a wild ride. In the year 1918 there were five known convex pentagons which could tile. Three more were discovered over the years until 1975, when the mathematics writer Martin Gardner reported that these eight pentagonal tiling patterns were the only ones possible. Then, within six months, another one was found. It seems the proof which had been claimed was not as exhaustive as everyone had hoped, and we now know they were only about halfway.

In the following years, some more pentagons were found by professional mathematicians and some were found by amateurs. In the late 1970s and 1980s, artist and recreational mathematician Marjorie Rice found four pentagonal tiling patterns which professional mathematicians had missed. Marjorie found her tiling patterns by drawing them out on index cards on her kitchen table. By 1985 there were fourteen known ways convex pentagons could tile a surface, and then all went quiet. Time passed. We celebrated the new millennium. I wrote a book with a whole section on tiling patterns (which I incorrectly assumed were now set in stone). Then in 2015 a new tiling pattern dropped!

This time it was found by computer. Professors of mathematics Jennifer McLoud-Mann and Casey Mann at the University of Washington Bothell (along with then-undergraduate David Von Derau) did the clever coding required to track down a tiling pentagon everyone had been missing for the previous 30 years.

Finally in 2017 even more coding occurred and French mathematician Michaël Rao was able to find and check all 371 different ways pentagons could meet at a corner, and confirm that there were only fifteen different ways that a tiling was possible. The Washington Bothell crew were also working on a similar exhaustive search, but were pipped at the pentagon. In theory, this completes the discovery and categorization of all possible convex polygon shapes which tile the plane by themselves.

I feel compelled to add that Rao's proof is believed to be correct but at the time of writing, a good seven years after it was released, the complete confirmation is still being conducted. It is no easy task to check such a complex computer-proof. So I wish to flag up that mathematicians are currently only 99.9 per cent sure there is not a sixteenth tiling pattern we've missed.

Armed with my copious research, I finally spoke to the builders. It turns out squares and rectangles are the only off-the-shelf shapes the paving stone comes in, and every extra cut through the stone comes at a cost of time, effort and builder goodwill. But if I could combine the pre-cut squares with a second easy-to-cut shape, then the project was still feasible. I sadly put away all my research into single-shape patterns but, on the upside, I do have a favourite two-shape tiling: snub-square tiling. It only requires squares and equilateral triangles. I decided to combine the pairs of triangles into rhombuses to simplify the cutting and paving somewhat. A mere three cuts through a rectangle, and the rhombus was ready to go.

This pattern is based on two of the only tiling shapes which have all sides and angles the same: the equilateral

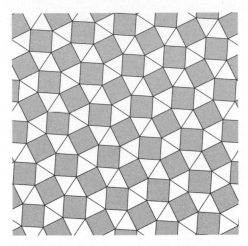

I hope this pattern doesn't snub you the wrong way.

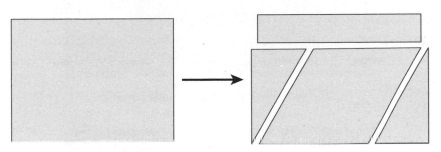

An acceptable amount of faffing about with a stone-cutting saw.

triangle and the square. Which was some compensation for wasting all my research into convex polygons, but I lamented that I wasn't able to incorporate the third of only three such 'regular' tiling polygons: our old friend the hexagon. That is, until I looked at the offcuts from making the rhombuses and something occurred to me.

Six equilateral triangles can combine to make a regular hexagon because their angles match nicely: the angles in a hexagon are 120° and equilateral triangles are all 60°. The skinny rectangle offcuts were not salvageable but the left-over triangles had angles of 60° and 30° which I could use to make hexagons. I sketched out a pattern for the builders to combine the offcuts to pave a walkway down the side of the house. If you look at it for a while you can see sections of hexagons hiding in plain sight.

A photo containing much labour and labrador.

The main course now comes with a side of hexagons.

Pack It Up

In 2018 I hosted the launch of a major vacuum cleaner company's latest model. They had a lot of stats and figures about battery life and whatnot to be conveyed, and it is apparently easier to teach a mathematician how to dismantle a vacuum cleaner while talking on stage than it is to teach a vacuum salesperson how to memorize and recite numbers in a coherent manner.

While chatting to the engineers, I discovered that they were also in charge of designing the packaging. I had always assumed that once the product was designed it was then passed off to a different packaging team, but no: that was done hand-in-hand with the electrical and mechanical engineering. 'That is a lot of effort for a cuboid!' I thought. (Mathematicians use the suffix '-oid' to mean any similar shape in the same family, but with more relaxed requirements. A cube has all sides the same length whereas a cuboid (strictly speaking, a rectangular cuboid) is any box even if the edge lengths vary.)

What I'd failed to appreciate was the supply chain. These machines would be manufactured in one country and then shipped to the different markets where they would be sold. That shipping is not cheap, or indeed that great for the environment. Making the product as convenient to ship as possible basically came down to one factor: how many could fit in a shipping container.

Shipping containers come in very standard sizes. Most of them are 8 feet wide, 8.5 feet tall and some multiple of 10 feet long. Just looking at the width to start with: 8 feet is 244 centimetres, so if you had five boxes exactly 48.8 centimetres wide they would fit in perfectly. However, if a company was not thinking about shipping containers, and unwittingly made their boxes a neat 50 centimetres wide, only four of them would fit, with the fifth box being just barely too big. By shaving 1.2 centimetres off the width of the packaging, this hypothetical company could fit 25 per cent more products in the same size shipping container. And that's just in one dimension!

Engineers are acutely aware that the final packaging needs to be a neat fraction of the dimensions of a shipping container. But that can be a real balancing act. A vacuum cleaner comes as a number of solid parts which each need a certain amount of space. Making them fit into a cuboid which tiles neatly to fill an exact shipping-container space is like a complex Tetris dance, one a company will use their top engineers to solve like they would any other engineering problem. The next time you use your vacuum cleaner, take a moment to appreciate that the precise size of the handle, hose and canister might be because in the 1950s someone thought 8 feet was a good size for a shipping container.

I asked James Bull, the head of packaging at supermarket chain Tesco, if they also alter the size of their products to make shipping them easier. The answer was a resounding yes, but instead of shipping containers Tesco worry about pallets and shelves. The Tesco packaging guru explained they need to make products which are both a neat fraction of a pallet and fit on as many shelves as possible. James expressed extreme envy for the newer supermarkets like Lidl with their modern, custom-built shops and standard shelf sizes. Tesco has a lot of old legacy shops with arbitrary shelf sizes, which makes it a nightmare to design products to fit.

We were specifically talking about cheese packaging. A shopper named Adam had bought cheese from Tesco and noticed that the packaging claimed to be a 41 per cent reduction in plastic on the previous packaging. Adam contacted my problem-solving podcast *A Problem Squared* to see if I could verify that this was geometrically possible. I ran the numbers and found it was extremely unlikely that a 41 per cent reduction was possible by merely reshaping a cuboid of cheese. Suspecting some kind of cheese cover-up, I went right to the top.

James explained that yes, the change in cuboid (my words, not theirs) was part of the plastic saving. There were also some other structural changes, like removing a zip-lock feature their research had revealed not enough people were using, but there was a second impact from changing the cuboid: reducing the maximum span. This new shape had a smaller maximum length, and it is the maximum length which determines the thickness of the packaging. If the span is bigger, the packaging is under greater stretching and tearing forces, and so thicker plastic is needed.

But that thickness cannot be reduced too much. Once again, it all comes down to shipping. The packaging itself only contributes about 10 per cent of the carbon produced to make and sell the product. Which means there is a trade-off between reducing the packaging and reducing how much food gets spoiled in shipment. A reduction in the thickness of packaging could easily offset any possible carbon gains by increasing how much stock spoils in transit and cannot be eaten.

It occurred to me that this freedom to change vacuum cleaner boxes and cheese packaging so freely is because any cuboid can tile 3D space. Like the 2D polygons, they are 3D polyhedrons with straight edges and flat faces. Some polyhedrons fit neatly together with no gaps and some don't. Much like every quadrilateral tiles in 2D space, every cuboid tiles nicely in 3D. Package designers don't even have to worry about its stackability when they tweak the exact dimensions of a box. This made me wonder if maybe there were more exotic polyhedrons designers could use instead. I very much doubted I could find anything more practical than a cuboid but I was certain I could uncover something much more fun.

Any shape which tiles in 2D can be turned into a prism which tiles in 3D. In one sense, the prism is the most simple of all the ways you can make a 3D shape. Just take two copies of any 2D shape you have lying around and join them together into a 'tube' with a bunch of rectangles. All triangle and quadrilateral prisms tile, as will any prism made from the families of pentagons and hexagons we met before. But I find polyhedrons which are not prisms far more exciting; they're not just rehashing something which worked in 2D, they're doing something all new in 3D.

A tetrahedron is the 3D version of a triangle (think three triangles attached to a triangle base) but sadly not all tetrahedrons can be stacked in 3D. Including the regular tetrahedron with all edges the same length. It is a great cruelty that the universe lets the humble equilateral triangle tile so perfectly, but then denies the equilateral tetrahedron the same ability. You can be forgiven for thinking it should work; no less than the great Aristotle incorrectly wrote about how the regular tetrahedron can be repeated to fill space. (Both Ken Perlin's tetrahedron lattice and HP's colour-conversion tetrahedrons were irregular ones with unequal sides.)

I'm also going to discount tiling patterns which require more than one type of shape, even though the most efficient area-to-volume packing ever found by humans is a two-shape tiling called the Weaire-Phelan structure, discovered in 1994. No one wants to have multiple shapes of the same packet of cheese which need to be interlocked correctly to be able to stack. Plus two-shape tilings are lazy: you can arrange any shape in a repeating layout and then invent a second shape

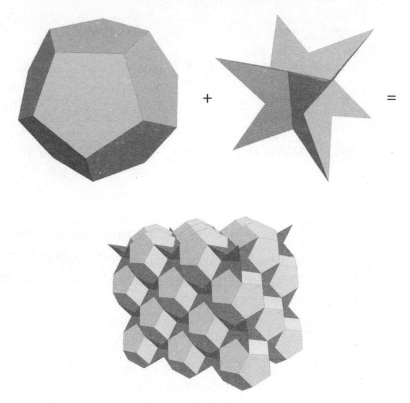

+ =

The endododecahedron is an inside joke.

which exactly fills the gaps. Even the regular dodecahedron can tile if combined with a shape called the endododecahedron ('endo' from the Greek meaning 'within').

For my ultimate 3D-filling shape I want a single convex polyhedron with all edges the same length. I'm going to give you two choices for your favourite, both of which have a claim for the title of '3D version of a hexagon'. It's the Rhombic Dodecahedron versus the Truncated Octahedron in a face-to-face match up.

The rhombic dodecahedron is made from twelve identical rhombuses and it fills space in a very gratifying manner with no need for an endo-buddy. I know you all have your

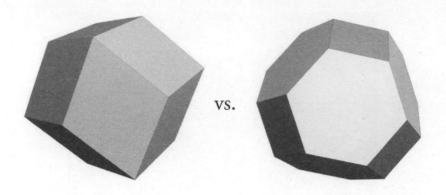

vs.

favourite dodecahedron and I think the rhombic is mine. It has a great hexagonal cross section. In the same way that if a bunch of tubes are stacked and squashed they form hexagons (like what happens in a beehive), if a collection of spheres are stacked and squashed you get rhombic dodecahedrons.

I particularly like it because all its faces are the same and opposite pairs are parallel. Which makes it ideal for construction projects. When I joined Adam Savage (of *Mythbusters* fame) in his workshop to build something fun for his You-Tube channel *Tested*, our only constraint was that it had to be completed in a single day. I had seen these art installations which use one-way mirrors and light to make the inside of the shape look like it extends out in all directions. But the more arty ones used a shape like the regular dodecahedron which, while fancy, does not tile space. I wanted to use a shape which tiles with itself so each reflected copy would align perfectly with all the others. We decided to design an infinite void lamp out of laser-cut acrylic rhombuses. Looking into it is a peek into an infinite lattice, vanishing in all directions.

The truncated octahedron has more than one type of face: it is assembled from eight regular hexagons and six squares. That's not as tidy as the rhombic dodecahedron with

Why use a cuboid when a rhombic dodecahedron is more terrifying?

its twelve identical faces. But if you look at the corners of the rhombic dodecahedron you'll see that there are two types: some are made from three rhombuses and some four. All the corners of the truncated octahedron are identical; they are all made from two hexagons and one square. Each shape is uniform in a different way.

The truncated octahedron may also edge out the rhombic dodecahedron because it uses less surface area for the same volume. That is to say, less plastic would be needed to wrap the same amount of cheese.

This problem was originally raised by William Thomson (aka Lord Kelvin) back in 1887. He framed it in terms of soap-bubble foam rather than cheese, but he had the same goal: to stack shapes with no gaps to maximize volume and minimize surface area. Thanks to Thomson's obsessive habit of writing down his ideas in notebooks, we know the problem first occurred to him while lying in bed on the morning of 20 September and he had found the truncated octahedron by 4 November.

Thomson's solution of the truncated octahedron is yet to be beat. Yes, the two-shape Weaire-Phelan structure would use 0.3 per cent less packaging, but for a single monotile in 3D the truncated octahedron is the current reigning champion. That said, mathematicians have yet to prove it is the best solution possible; for now, there is still the possibility that there is a better polyhedron out there, waiting to be discovered.

I think the truncated octahedron also has the best shot at emulating the ultimate killer feature of the cuboid: stacking with nice flat sides. Whatever shape we choose for cheese packaging, there's no denying it's going to have to live in a cuboid's world. Shipping containers, pallets and shelves are all themselves cuboids, which gives cuboid packing a distinct advantage. Truncated octahedrons get close to this perfection because they can be arranged in a cube and the outer faces will sit flush if they are bundled into a surrounding box. A few small gaps around the edges, yes, but they use 88.6 per cent of the wrapping area of cubes. Which is maybe not enough to upend the packaging world.

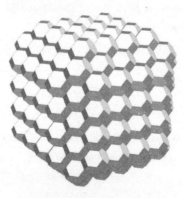

Lattice agree this is quite nice.

Part Bee

The classic picture of honeycomb is a lattice of hexagons, but those are only the openings to the cells. The front doors. There needs to be a back wall as well. Not only do bees stack their cells side-by-side and row-upon-row to share walls, they also build them back-to-back, neatly staggered so that each one can stick out and mesh with the ones behind. It seems bees have to solve the 3D tiling problem as well.

The same conditions apply as before: they want to save space and save wax. Given that bees have found the optimal solution of hexagons in 2D, surely they have cracked the 3D case as well and can adjudicate in the rhombic dodecahedron versus truncated octahedron debate? Well, bees have definitely picked a side: the end of cells in a beehive is ... a rhombic dodecahedron.

That's me on the right, checking beehive rhombuses with my own eyes.
Yes, I brought a cardboard model.

With a hexagonal cross-section, it feels like a perfect shape to end a hexagon tube. I had a look at some beeswax with a beekeeping friend, and not only did I get to wear the full bee suit but I could see where the hexagons give way to three rhombuses ending the tube. But is it the optimal shape? Are bees little, buzzy mathematical geniuses?

No. In 1964, Hungarian mathematician L. Fejes Tóth published a paper titled 'What the Bees Know and What They Do Not Know', and he comes out of the gate swinging. Here is the first sentence: 'In the first part of this paper we construct a more economical honeycomb than that of the hive bees for any parameters involved in the problem.' That is a serious throw-down on the bees, claiming humans can do better.

And we can. Tóth pointed out that with a few tweaks a truncated octahedron is a better shape to use than a rhombic dodecahedron. If it is distorted a bit, by squashing the hexagons and messing with the squares, it is possible to give it a regular hexagon cross-section shape which means it will fit nicely on the end of a beehive hexagonal cell. Plus, it will mesh perfectly with the layer of other cells behind it, all while providing the same amount of volume as a rhombic dodecahedron with 0.14 per cent less wax.

The existence of a superior shape to the rhombic dodecahedron rules out the possibility that bees are somehow divinely inspired to find optimal solutions to maths problems. They have merely evolved a good-enough solution, and the only-very-slightly better solution is sufficiently different that bees have not inadvertently stumbled across it.

And it is worth remembering that bees are not even trying to make rhombic dodecahedrons. They are merely pushing wax around, which happens to lead to a rhombic end to their cells. Yes, this behaviour may have evolved specifically because it leads to such near-optimal results, but that does not mean the bees are doing geometry.

Vincent Gallo also did experiments on the ends of bee-hive cells. When building in isolation, away from any other cells, bees will push the wax end of their cells out until it forms a little spherical dome. And if bees are encouraged to make cells directly aligned with the opposite ones they happily build the worst possible ending to the cell: a flat wall. So the 'bees as geometers' theory is debunked. Yes, they have evolved to make very efficient honeycomb out of wax but no, they are not doing any maths.

Don't Make Me Repeat Myself

For a long time, the holy grail for mathematical tiling patterns was a polygon which could perfectly cover a surface but in a way which never repeats. One thing all the tiling patterns we've seen so far have in common is that they repeat periodically. I only had to give the builders a small diagram of snub-square tiling because once they got the pattern correct it could be repeated forever. Easy.

Mathematicians dreamed of a shape which sat right on the cusp between order and chaos. Some polygons are able to cover a surface with no gaps in a neat pattern, and others cannot fit together without gaps. But imagine a shape which brings both sides together: it cannot form a repeating pattern yet it can still cover a surface.

This mystical tiling pattern is called 'aperiodic'. A lot of tiles can form a 'non periodic' pattern: square tiles can be arranged with each row offset a different, irrational amount from the previous. Technically, this is a pattern which never repeats. But an aperiodic pattern involves the stronger condition that it is impossible to arrange the tiles in a periodic fashion. Square tiles could be knocked back into a periodic pattern and so they don't count.

The first set of aperiodic tiles was found in 1964, but it involved combining 20,426 different shapes of tiles together. By 1974 this had been reduced to a set of two shapes, called Penrose tiles, which were aperiodic as a team, but the search was still on for a monotile that could be aperiodic all on its own. This mysterious, hypothetical shape was often called the 'einstein' as a hilarious German-language pun on 'one stone'.

Even though mathematicians had yet to find an einstein tile, they did know some things about what it must be like (if it did exist). Recall Rao's proof from 2017 showing that all the convex pentagons which could tile had been found. This completed the search for all convex polygons, and every single one which could cover a surface did so in a nice periodic fashion. If there was an aperiodic monotile out there, it was not convex. It must have concave, sticking-in bits.

In 2010 an einstein was discovered! But it was a terrible shape. The Socolar-Taylor tile, named after its discoverers, was an aperiodic monotile but it wasn't contiguous. Several little disjointed pieces all made up, technically, 'one tile'. Having tiles each made from a collection of disparate parts did feel unsatisfactory. In a follow-up publication the discoverers described it as 'an einstein according to a reasonable definition'. Which is absolutely true. But both mathematicians and builders agreed that each tile being a solid piece was an even more reasonable definition.

Then in March 2023 it was found. The first ever aperiodic monotile. See if you can guess if it was someone messing around on their kitchen table at home or an advanced computer search! Answer to follow. I remember the release vividly: the news broke on 21 March and on 22 March I was due to give a public lecture at the Royal Society in London

0.544639

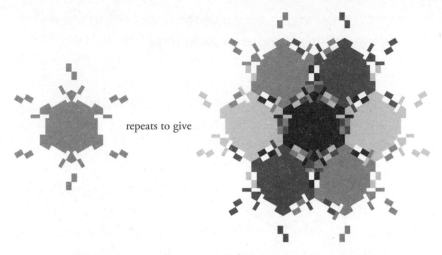

repeats to give

Do you think anyone would tile their bathroom with this? I grout it.

called 'Every Interesting Bit of Maths Ever'. A swift rewrite ensued.

The excitement was instant. It swept through the maths world very quickly, and the mainstream media was not far behind. The mathematicians who had found the shape had dubbed it 'the Hat' because they thought it looked like a hat. It has also been claimed to look a lot like a shirt. The point is, it was a nice, tidy, public-friendly shape. Before long people were 3D printing them, baking cookies shaped like them. My friend Ayliean MacDonald showed up for my Royal Society lecture in a Hat-covered dress she had made herself.

There was something about the Hat which made it popular with the public and mathematicians alike: it was surprisingly simple. Given this shape had been eluding the entire mathematics community for over half a century, nobody expected it to be so straightforward. It's a 13-sided polygon, far fewer sides than I would have predicted. It is concave, as expected, but doesn't have any detached, fragmented bits or any holes. When I look at it, I see a slightly modified equilateral triangle.

Even in the research paper announcing its discovery says, 'The shape is almost mundane in its simplicity.'

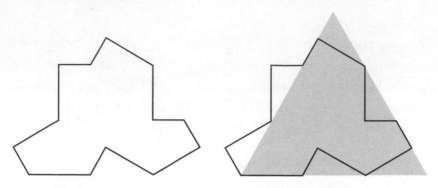

The Hat, which I see as an equilateral triangle with four bits cut off and two added.

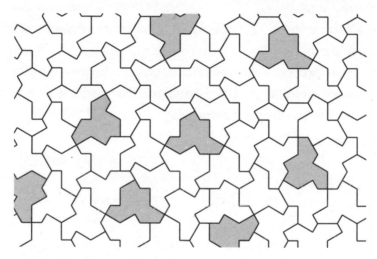

The Hat tiling (with flipped tiles in grey). It's insultingly simple.

None of this is to devalue the incredible feat of finding the Hat. It was discovered by a retired print technician, David Smith, doing some recreational maths at home on his kitchen table. He had been designing shapes in a tiling software package, when he outlined the Hat and realized there was no obvious way to arrange it in a tiling pattern. But it looked like it should fit together nicely. David cut 30 of them out of cardboard and found they did fit together but with no obvious pattern. Another 30 copies were cut out and added to the tiling; still no pattern.

He contacted mathematician Craig Kaplan, who used some adapted software to explore how far the Hat could tile. It tiled further than any other known non-tiling shape, which strongly suggested the Hat could indeed cover any infinite surface. Yet the patterns it formed were not periodic. More mathematicians were recruited, and soon they managed to prove that the Hat was indeed an aperiodic monotile. For completeness they even proved it two different ways. The first proof was done using a computer, which worked but didn't offer any insight into why the shape was aperiodic. As they said in the paper, 'These calculations are necessarily ad hoc, and are essentially unenlightening.' So they proved it again in a much more satisfying way. There was now no doubt this was the einstein shape everyone had been looking for.

Then David found another one.

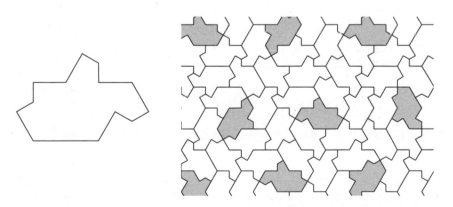

I think we can all agree this is the shape of a turtle wearing a hat.

Dubbed 'the Turtle', it was a second example of an einstein. It felt wildly unlikely that two unrelated einsteins would be found so close together by the same person. And, after a bit of digging, the tile team found that the Hat and the Turtle were in fact two members of the same 'family' of tiles. This is the same as how we consider all rectangles part of the same family of shapes, each member of the family having a different ratio between the two edge lengths. Actually, since the ratio in a rectangle can be anything, the family of rectangles is infinite. The same is true of the Hat family, but it's a less straightforward ratio. The original Hat is made from two different side-lengths (1 and √3) and those lengths can be varied to produce other einsteins.

When the side-lengths are the other way around, √3 and 1, the resulting shape is the Turtle. All the other ratios work as well, with three exceptions. If the entire infinite family of Hat tiles were put in a line, and labelled with their distinguishing two side-lengths, it would start with tile 0, 1 and end with tile 1, 0. Both of those end tiles are not technically aperiodic. They can be arranged to be nonperiodic but also have alternate periodic arrangements.

Strangely, the very middle 1, 1 tile is also not aperiodic. For a shape to be aperiodic it needs to walk a very fine line between order and chaos: too much order and it becomes periodic; too much chaos and it ceases to completely cover a surface. Having the two edge lengths the same tips this middle case into having just enough order to be periodic. But, on the plus side, we are still left with infinitely many other shapes which do work.

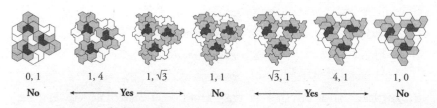

A few example members of the Hat family.

Classic maths. You wait half a century for one aperiodic monotile and then infinitely many of them show up at once. The only slight disappointment was that all of these tilings use the reflection of the tile within the tiling. Which is something that mathematicians are OK with, but actual bathroom tiles and paving blocks come with a front and a back. So, annoyingly, the Hat would not make a good bathroom tile. For that a new einstein will need to be found which tiles without using its reflection. We can only hope.

And that hope has already paid off! In May 2023 the same team came back with a chiral aperiodic monotile – one that tiles without using reflections – just over two months after the first einstein had been announced. I will add that two months was the perfect amount of time for the mathematics-communication community to have just finished work on all manner of podcasts, videos, blog posts and magazine articles telling the 'definitive' story of the einstein before bam: all obsolete. (Goodness knows what will be announced the second this book gets published.)

This new shape was named 'the Spectre', and it was also hiding in plain sight. David found it right in the middle of the Hat family: it's the shape with 1, 1 edges that we had previously discounted! All of the Hat tilings which were aperiodic needed to use their reflections, but it was the reflected versions of the 1, 1 tile which stopped it from being aperiodic. If reflections were banned then it would become aperiodic. David and the team realized that by curving the edges in a special way they could remove the ability for the reflected version to fit at all, turning the Spectre into a 'strictly chiral aperiodic monotile'. Mission accomplished!

I feel like, over time, the general public gradually builds up the capacity to pay attention to a breaking maths story (like a video-game power bar), and the Hat came out at just the

0.469472

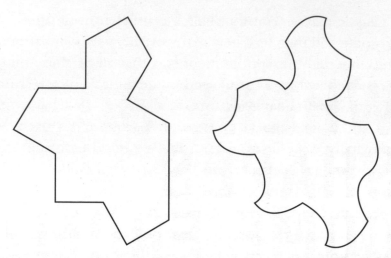

1, 1 tile on the left and disguised as the Spectre on the right.

right time, depleting the reservoir of excitement. When the Spectre was announced two months later it didn't even register as a blip on the mainstream media or in the public consciousness. Sure, maths people were super excited – this was arguably the more amazing result – but the general populace had no need for another new shape so soon. Even though this one is ideal for tiling a bathroom.

At the time of writing I am wondering what the next startling new shape will be. It could come from anywhere. I have contacted the Hat team to double-check they don't have some other new tilings to be announced the moment I finalize this manuscript. Because after they found the whole Hat family and saw how non-exotic the shapes were, they wrote, 'we might therefore hope that a zoo of interesting new monotiles will emerge in its wake'.

I also hope it does. But not until the next edition of this book.

Six

WHERE DO SHAPES COME FROM?

The Rising Star cave system in South Africa contains some tight squeezes; to reach one particular collection of chambers, 120 metres from the nearest cave entrance, visitors need to pass through gaps less than 20 centimetres across. The lead paleoanthropologist who first explored them was so keen to get down there they went on a diet and lost 25 kilograms just to fit though (and tore a rotator cuff on the clamber back out). It would have been an easy fit for some earlier visitors, the now-extinct species of smaller-stature humans *Homo naledi*. They seem to have visited these deep cave chambers quite a lot, and left behind what is believed to be the earliest known human art.

Roughly 300,000 years ago, as we *Homo sapiens* were considering evolving into existence, and well before human language had evolved, these ancient cousins of ours picked up rocks and laboriously scratched markings into the hard cave walls. Markings that featured geometric patterns. Markings that include triangles. Yes, the earliest known human doodling, from over a quarter of a million years ago, was a triangle.

Does this count as the invention of the triangle? I don't think trying to answer that question is worth a complete detour, but I will say that I find it hard to convince myself that the notion of 'triangleness' did not exist in the universe until a sentient animal happened to scratch one into the rock wall of an inconvenient cave. So I'm just going to declare this book is in the 'maths is discovered' camp and not answer any further questions on the matter.

Now we can imagine the universe has a theoretical storeroom containing all of the shapes. Every now and then, humans will unearth a new one. The triangle was an 'easy' early discovery, and the flow of new shapes has not slowed down. More than one new shape has been discovered while I have been writing this book, which makes writing it feel like a game of whack-a-gon.

So what counts as a shape? You could grab a piece of paper right now, scribble some semi-arbitrary lines and curves and declare your result a new shape. It is undoubtedly true that no one else has drawn the same set of franken-lines as you. But I am going to upgrade the universe's storeroom to be a discerning collection of 'proper shapes'. A proper shape is a shape which has at least one of these qualities of distinctness: it is somehow unique, conforms to some interesting constraints, solves a practical problem or is a mascot for a whole family of shapes. The universe's store house now has some level of quality control.

The constant drip of new shapes being discovered by mathematicians (and maths enthusiasts) occasionally reaches the attention of the wider public. Like when in July 2018 some mathematicians in Spain announced a new shape they called 'the scutoid'. Thanks to some combination of the shape being media-friendly and it being a slow news day, the scutoid hit the popular press and quickly elicited the classic response: disbelief and incredulity that there could still be new shapes to discover.

The new shape was the result of a group of biologists coming up against a problem we have seen before: packing shapes together without any gaps. Their results were titled 'Scutoids Are a Geometrical Solution to Three-Dimensional Packing of Epithelia'. The only biology-centric word we need to unpack there is 'epithelia' and, to quote the first line of the paper, 'Epithelial cells are the building blocks of metazoa.' Thanks, biologists, for clearing up that one word we didn't understand with another one we don't understand. Like me saying 'Scutoids are just a type of prismatoid.'

It transpires that 'metazoa' is a fancy way of saying 'animals'. Not sure why they didn't just say that. The point is, it's us. Epithelial cells form in layers and make up the outer surface in everything from our skin and eyeballs to our lungs and pretty much any body cavity you care to think of. The biologists wanted to know exactly what shape these epithelial cells were, that allowed them to pack together in three dimensions.

I've previously been dismissive of prisms as a way to fill 3D space, but that is because I am looking for things which are mathematically interesting. Epithelial cells, like bees, do not care about being interesting. They have merely evolved to fill space in an efficient manner and prisms are a good way to do that. Prisms have a shape at each end which fits together in 2D and then those surfaces are joined up with rectangles to make a 3D shape. It feels like an obvious way that these cells could stack together to form a layer of tissue.

Prisms on their own, though, don't explain the shape and curvature the biologists had observed in the structures these cells would form. The layers of epithelial cells were not flat, they would curve about in a way that perfect prisms – which have parallel sides with the same shape at either end – could not explain.

Like a cuboid is a more general cube with different length sides, a prismatoid is a prism with fewer constraints. If the face at one end is smaller than the other, that type of prismatoid is called a 'frustum'; it's a prism which tapers, or you can think of it as a pyramid with the top cut off. And, far from being considered pointless, this is what biologists had thought enabled layers of epithelial cells to take on curved shapes.

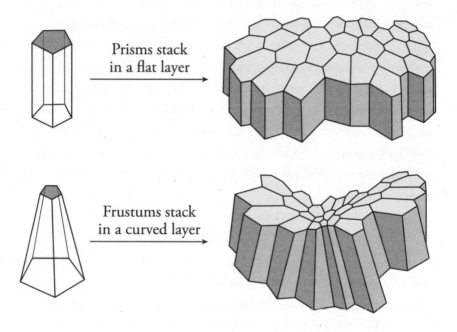

Prisms stack in a flat layer

Frustums stack in a curved layer

These layers of cells are a bit messier than tiling we've seen before, as all the individual cells are different shapes. The goal of the biologists was to categorize what types of shapes might be joined together to make these layers, and their reasoning was that the cell surfaces on one side get smaller and so prism-shaped cells would become frustum-shaped as the surface curved. But that alone could not explain what was seen in actual animal tissues. They were missing a novel-shaped piece of the puzzle.

The next member of the prism family occurs when a face gets rotated instead of shrunk. An 'antiprism' is when one face is rotated to be out of alignment with the opposing face, such that each of its corners sits at the midpoint of the other shape's corners. Then, instead of joining the two polygons with rectangles, we use our good friend the triangle. This use of triangles makes for a much more exciting shape! We've already seen that boring modern architecture has given us endless cuboid buildings. Sure, we could try frustum- and pyramid-shaped buildings to spice things up, but that is so four-millennia-ago. The real interesting move would be antiprism buildings. And one very big example has already been built!

The tallest building in the USA is One World Trade Center in New York City's rebuilt financial district. It stands 1,776 feet tall, some of which is a cuboid-base and some of which is the sculpted antenna on top. The bulk of the building is a square antiprism. The building starts with a square shape and then eight glass triangles reach skyward: four from the edges and one from each corner. They eventually reach the top square which is rotated 45° from the base. This means that every floor in between is an octagon, with the middle floor being a regular octagon with equal-length sides.

One World Trade Center tapers as it gets taller (it's an antiprism and a frustum; an antifrustum). It's a bit hard to see clearly from the ground, so when I was in NYC with my friend Laura Taalman she 3D printed a model of the building so we could see the geometry a bit more clearly. Classic Laura. The top of the building is exactly sized so that, despite being rotated by 45°, the corners do not stick out.

Before the trip we had been discussing what the volume

0.374607

of this kind of antifrustum would be, and the calculation seemed to be rather complicated. Then Laura had the insight that, because of the specific size of the top of the building, the volume of the WTC antifrustum is equal to the volume of a cuboid with the same base, minus the volume of a pyramid with a base the same size as the top of the building. I appreciate that is one heck of a sentence and I couldn't get my head around it until Laura demonstrated it with the models. Is it a very pleasing and potentially unnecessary bit of geometry? Yes. Does it indicate that the world is run by a shadowy group of geometers? No comment.

In the search for the shape of epithelial cells, after antiprisms (and antifrustums) we have all the mix-and-match options called 'prismatoids'. This is where two different flat polygons are positioned parallel to each other and then connected up using all manner of rectangles, trapezoids and triangles. These came into play because scientists were investigating epithelial cells which had formed a tube. They knew each cell must have a face on both sides of the tube wall, and the arrangement the cells form is fairly well understood (they follow a Voronoi arrangement, for anyone familiar with such

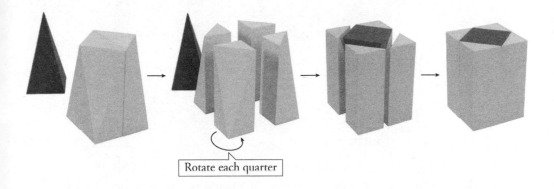

Rotate each quarter

The tower plus a pyramid is equal to a cuboid with the same footprint.

things). They could run a computer model starting with the central points of the inside and outside faces of each cell. They found that the way the cells connect to their neighbours changes on each side of the surface.

On one side of the surface a cell might contact five surrounding cells, but on the other side it touches six! Cells were not simply getting bigger or smaller across the layer, but they were changing how many neighbours they had. This implies a shape which has a pentagon at one end and a hexagon at the other end, which can be achieved by slipping one triangle into the prism.

Except it didn't seem to work quite like that. I spoke to one of the mathematicians brought in to help the biologists

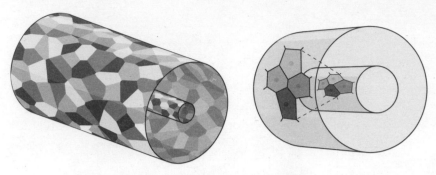

A complete tube of cells, and just a few cells where the outside surface is matched to the inside one.

(my words, not theirs), Clara Grima, who iterated a mathematical model until it eventually matched what could be seen under the microscope. This is a common back-and-forth in science, where mathematicians come up with a hypothetical maths explanation for how reality might work and use that model to make a prediction, and then observational scientists check to see if that is what they actually see. Using a full-length triangle on the side of the cells didn't exactly match what biologists saw manifesting in the behaviour of the cells. A few iterations later, they determined that the shape of the cells was something all-new which used a Y shape to join two vertices into one. They named it the scutoid.

This means I misspoke earlier, when I was being deliberately obtuse: scutoids are actually not a type of prismatoid, despite being very closely related. Prismatoids have all of their vertices on the ends in two parallel planes. The scutoid has that extra, rogue vertex hiding in the middle; a vertex that spent aeons in the middle of layers of epithelial cells, out of sight of trying biologist eyes. It was only revealed by a clever mathematical investigation.

There is one other feature that distinguishes scutoids from prismatoids: their faces are not flat. My friend Laura

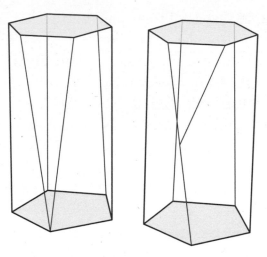

*A lazy prismatoid to go from hexagon to pentagon, and the slightly
more exciting solution nature actually uses.*

Taalman found this when she was the first person to try and
3D print some scutoids and the resulting shapes did not fit
snugly together, as we'd expect the actual cells to do. She had
to adjust her code to allow for curved surfaces, and then it all
worked. This means, as shapes, scutoids are among the more
all-encompassing 'solids', rather than the stricter 'polyhedrons'. You can find Laura's STL files online (2,208 triangles
per scutoid).

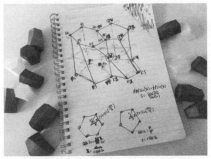

Laura's 3D-printed scutoids, and the page of maths it took to make them work.

The scutoid managed to swing from technical, scientific publications, to popular-science outlets, to the full-on mainstream media. It's not often UK tabloid the *Mirror* reports on a shape with no mirror symmetry, but their headline was quite straightforward: 'Brand New "Scutoid" Shape Discovered by Scientists in Skin Cells'. The *New York Post* went with 'Scientists Find Brand New Shape Living Inside Us All', which is a bit ominous, and Forbes had the more sober, but still quite funny, 'What Is a Scutoid? It Is a New Shape Maybe All Over You'.

Nobody used my suggestion, 'The New Scutoid Shape Will Grow On You'.

Do Me a Solid

There are three names in the shape-finding business which rise above all others: Plato, Archimedes and Johnson. Each one has a whole collection of 3D polyhedrons named after them, and they were all in search of shape perfection.

Perfection for shapes means keeping things consistent. We've already seen equilateral triangles, where 'equilateral' means that all the edges are the same length, and other regular 2D shapes, where all the lengths are the same *and* all the corners are identical. Plato, Archimedes and Johnson each tried in their own ways to see how far they could take these ideals in 3D.

Plato lived from 427 to 347 BCE and was a bit of a big deal as far as philosophers in ancient Athens go. I appreciate that trying to summarize all of his intellectual thought into one sentence will anger a lot of people, but his whole thing was that our inaccurate mathematical scribbles, in our messy reality, are only approximations of the true, beautiful mathematical ideals. He was all about how neat and tidy mathematics was in the abstract. Which makes it logical that the most regular and pure 3D shapes are named the Platonic solids (the name 'solid', which is a more catch-all version of 'polyhedron', is not needed here, but is used because of historical inertia).

Plato was 'first to market' and he picked off all of the most regular shapes. Regular polygons are the self-nominated public faces of 2D shapes; when we say 'pentagon' or 'heptagon', what most people imagine is a neat and tidy regular pentagon or regular heptagon (when in reality a pentagon can be a shape with any haphazard collection of five sides of any lengths). There are infinitely many 2D regular polygons but there are only five 3D regular polyhedrons, and Plato claimed them all.

With three dimensions, forming a regular shape is slightly more complicated. As well as each 2D face having identical corners, there are now the additional corners on the 3D shapes where the faces meet. To avoid ambiguity, these 3D corners are called 'vertices'. The Platonic solids are the five shapes where 2D regular polygons can be combined into even-more-regular 3D polyhedrons, which now also have identical vertices. Triangles are the winner here, taking home three of the five possible arrangements: four equilateral triangles make the tetrahedron; eight, the octahedron; and twenty, the icosahedron. Lagging behind, six squares manage to make a cube and twelve pentagons will result in the regular dodecahedron.

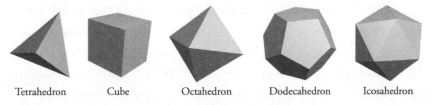

Tetrahedron   Cube   Octahedron   Dodecahedron   Icosahedron

The Platonic family photo.

Interestingly, Plato did not actually discover these shapes, nor was he even the person to prove that there are only five. There are examples of humans making these shapes (or objects we have since convinced ourselves are these shapes) going back a very long way before Plato, and it seems Pythagoras could have studied all five of them a century earlier. And it was after Plato, in Euclid's *Elements* written around 300 BCE, that they were put on a firm theoretical founding. Ironically, it seems Euclid's work might be based on Plato's contemporary and buddy Theaetetus. It turns out everyone but Plato was involved!

What Plato did do was write about the shapes in a very compelling way. In his work *Timaeus* he outlines how each of the elements were made of 'atoms' shaped like regular polyhedrons. Earth was made of very tiny cubes; fire was tetrahedrons (spiky!); air, octahedrons; water, icosahedrons; and the whole cosmos itself was a big ol' dodecahedron. This was enough for Plato's name to be forever linked with these ideal shapes, though we now suspect his friend Theaetetus did the mathematical heavy lifting.

If you're ever in Athens and want to pay your respects to these mathematical forebears, the land where Plato's academy stood is now a public park. In his time it was a walled park with statues and temples, used for sporting activities, festivals and generally hanging out. Plato lived nearby and had a garden in the park he used for his teaching. It was Plato setting up shop on land named after someone called Academus which gives us the modern word 'academy' for a place of learning. I dropped by once, and found a section of ruins which was most likely part of the complex where both Plato and Theaetetus sat around to chat about shapes. Now it is just a collection of stones which, I'm sad to report, are mere cuboids.

Where the building blocks of Platonic solids were first laid.

Next up is Archimedes who lived from 287 to 212 BCE, safely after the confusion of the Platonic solids. From Sicily, Archimedes did all sorts of incredible mathematics, including being the first person to really pin down a semi-exact value of pi. He narrowed it down to within $3\frac{1}{7}$ and $3\frac{10}{71}$, which is pretty impressive for the time period. Plato had nabbed all of the shapes with identical faces and identical vertices. To get the next family of shapes there is a choice about which restriction to keep: consistent faces or consistent vertices.

The Archimedean solids ignore the faces and focus on the corners: all of the vertices still have to be the same, but it's now permissible to mix and match the regular polygons which make them up. No longer do equilateral triangles and squares have to remain in their separate solids, they can join forces in the cuboctahedron, the rhombicuboctahedron and the snub cube. All in, we get thirteen new shapes with all the regularity of the Platonic solids, but we no longer care if the faces are indistinguishable. I like the Archimedean shapes because they have a bit more spice and flavour than the Platonic solids, yet are still pleasingly symmetric. They also include the truncated octahedron, which we've already met, and the truncated icosahedron, which is the classic football shape.

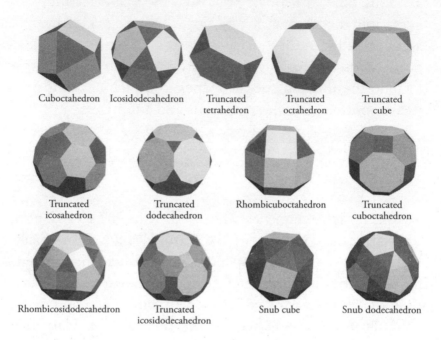

Cuboctahedron Icosidodecahedron Truncated tetrahedron Truncated octahedron Truncated cube

Truncated icosahedron Truncated dodecahedron Rhombicuboctahedron Truncated cuboctahedron

Rhombicosidodecahedron Truncated icosidodecahedron Snub cube Snub dodecahedron

The only issue with calling this group of thirteen the 'Archimedean solids' is that we are excluding infinitely many other shapes. Specifically, two families of shapes. Which is the result of these four facts:

- The Archimedean solids are made from combinations of regular polygons.
- There are infinitely many regular polygons.
- Any regular polygon can be made into a prism using squares.
- Any regular polygon can be made into an antiprism using equilateral triangles.

This means we can make as many Archimedean solids as we want, all with regular faces and identical vertices. The only technicality you can sweep them away with is 'there are too many of them'; anything else is mathematically disingenuous. Since the square prism and the triangular antiprism get to join the Platonic solids as the cube and octahedron, I'd argue it's a bit mean to exclude the rest from the Archimedean solids. So, technically, the Archimedean solids should include the thirteen special shapes, the Platonic solids and the two infinite families of prisms and antiprisms.

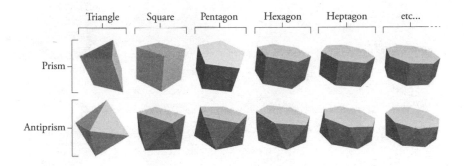

While we're expanding the family, I'd like you to meet Catalan solids. Catalan is the anti-Archimedes, answering the question: what if we did care about all of the faces being the same but didn't care about the vertices? The Catalan solids were catalogued by Eugène Charles Catalan in 1865 and they are a shadow society of shapes, which match up exactly with the Archimedean solids. The 'dual' of a polyhedron is the shape you get if you put a dot right in the middle of every face and then join those points up as the vertices of a new solid. If you do this for any Archimedean solid you'll get the matching Catalan solid.

You'll see some old friends here. The rhombic dodecahedron has shown up before, and our 120-sided dice, the disdyakis triacontahedron, is here to play. These shapes with

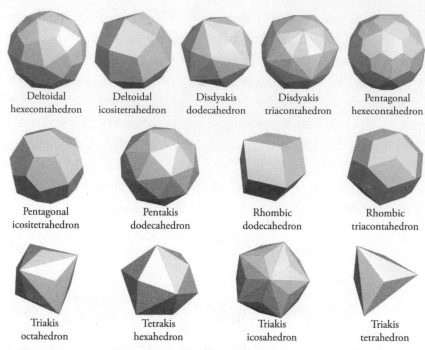

| Deltoidal hexecontahedron | Deltoidal icositetrahedron | Disdyakis dodecahedron | Disdyakis triacontahedron | Pentagonal hexecontahedron |

| Pentagonal icositetrahedron | Pentakis dodecahedron | Rhombic dodecahedron | Rhombic triacontahedron |

| Triakis octahedron | Tetrakis hexahedron | Triakis icosahedron | Triakis tetrahedron |

indistinguishable faces are called 'face-transitive' and that is why they make such good dice: if all faces are identical then they are equally likely to be rolled. In addition to the duals of the Archimedean solids, these face-transitive shapes also include the duals of the Platonic solids (which are ... the Platonic solids) and all of the duals of the regular prism and antiprism infinite families.

For completeness we end with the Johnson solids. After the titans of Plato and Archimedes, I appreciate it's a bit jarring to end with a mathematician named Norman Johnson from the 1960s but here we are. Norman completed our journey by looking at all the possible ways regular polygons could be formed into any polyhedrons with no further restrictions. In addition to the Platonic and Archimedean solids, in 1966 he added ninety-two more shapes (and in 1969 Victor Zalgaller proved Johnson had not missed any). I'm not going to show all ninety-two (to my mind, they get a bit chaotic) but here are a few fun ones.

Pentagonal pyramid: five equilateral triangles on a pentagon. Also friends with the square pyramid, but watch out: the triangular pyramid is actually just the tetrahedron which has already been claimed by Plato.

Gyrobifastigium: two triangular prisms with a twist.

Elongated pentagonal gyrocupolarotunda: this one is so dumb. It's a pentagonal cupola and a pentagonal 'rotunda' stuck on either side of a decagonal prism. I hate it, but it technically counts.

0.173648

Sphenocorona: one of the few Johnson solids which is not just a hack-job on other shapes. Plus its name means 'wedge crown', which is pretty cool.

That's 110 convex polyhedrons which can be made from regular polygons: the five Platonic solids, thirteen Archimedean solids and ninety-two Johnson solids. Plus the infinite families of regular prisms and antiprisms, and why not throw in the Catalan solids for completeness. Phew. At last we are done.

An Edge Case

If you think that is case closed, you are wrong! Although mathematics is seen as the subject where there are indisputable right and wrong answers, mathematicians love to argue over what they believe is correct. Which brings us to the elongated square gyrobicupola, a shape whose name is as annoying as its classification.

On the left you can see the rhombicuboctahedron, made from eighteen squares and eight equilateral triangles which all meet at vertices of three squares and one triangle. On the right you have the elongated square gyrobicupola, also made from eighteen squares and eight equilateral triangles which all meet at vertices of three squares and one triangle. The left

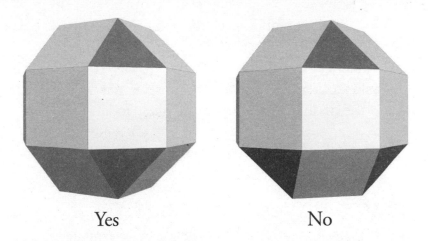

Yes   No

one is an Archimedean solid. What do you think about the
one on the right? Should it be Archimedean as well?

To turn a rhombicube into a gyrobicupola you just need to
grab one of the six 'squares surrounded by squares' and give
it a firm twist. It should click one step around, dragging its
adjacent squares and triangles with it. If you stare at the two
shapes long enough you'll notice on the left any close-by pair
of triangles flank a square, whereas on the right that is no
longer true. If you were to give me both of these shapes I'm
confident that after a close inspection I could tell them apart.
They are distinct shapes. If they both meet the Archimedean
criteria of having regular faces and identical vertices, why is
there any ambiguity?

It's because mathematicians get real fussy about what
'identical vertices' means. All the vertices on a gyrobicupola
may be the same in isolation, but if you look at what neighbours they have they fall into two different camps.

I think of it as comparing the rhombic dodecahedron
to a shape called the Bilinski dodecahedron, discovered in
1960 by Croatian mathematician Stanko Bilinski. It also has
twelve identical rhombus faces. But whereas all of the rhombic dodecahedron faces are indistinguishable even when

factoring their neighbours – the true definition of being face-transitive – the faces of the Bilinski dodecahedron are locally identical but globally distinguishable. You'd not want a Bilinski dodecahedron dice because it has a 'flat patch' around the middle which it is more likely to land on.

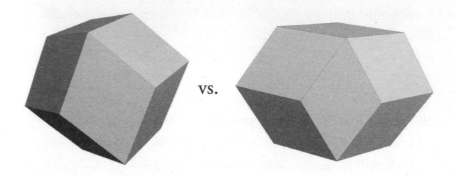

vs.

The general consensus is that the lack of global symmetry relegates the gyrobicupola to the Johnson solids, but there have been modern efforts to get it promoted back to Archimedean (most recently by mathematicians Branko Grünbaum, in 2009, and Katie Steckles, in 2018). The debate rages on. We cannot check what Archimedes himself had to say because, while we know Archimedes definitely found thirteen shapes, what he wrote about them is now lost to the ages. Many of Archimedes's writings have survived for us to read, but not this one crucial text. We know it exists because the author Pappus, writing sometime around the 300s BCE, mentions this work by Archimedes and that it contained all thirteen shapes. But we do not know what Archimedes's logic was for selecting those original thirteen.

I suspect that the number thirteen has become a self-fulfilling prophecy and we've honed our modern definition of an Archimedean shape so that the total count aligns with what Archimedes thought. It would be very interesting to compare our modern logic with the ancient version to see if we've achieved the same result via the same means, or if we've given it a twist.

Surely We Are Done Now

You would think, after millennia of looking, we'd have all of the regular polyhedrons found and neatly categorized. So it was a bit surprising when, in 2011, a fifteen-year-old found a new one.

In this case it was an equilateral pentagon dodecahedron. There is only the one regular pentagon dodecahedron with all identical vertices, but to be *equilateral* we only need the edge lengths to all be the same. We have seen one such equilateral pentagon dodecahedron already: the endodo-decahedron which filled the gaps between regular pentagon dodecahedrons so they could tile in 3D. It is a concave, pointy shape to fit into the gaps, but because it needs to mesh perfectly with the regular solids, it also has identical edge-lengths.

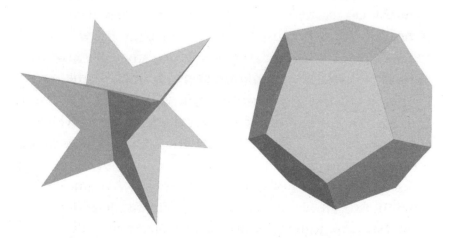

Innie and an outie.

This is exactly what fifteen-year-old Julian Ziegler Hunts found in 2011: yet another regular equilateral dodecahedron. He was messing around with some code to explore 3D shapes with ten faces, because he was about to go to the tenth Gathering 4 Gardner, a recreational maths conference I also frequent (no surprises there). Each year the attendees show off their latest bits of ridiculous maths, and there is an extra motivation to find things which pertain to the current conference number. Looking forward to future conferences, Julian ran his code on shapes with more faces and after a while his computer showed that there was an equilateral dodecahedron with a height of -1.0615284. Yes, that is a negative height.

The negative height is an artefact of defining the classic Platonic dodecahedron as zero, and positive heights equated to stretched shapes. It turns out that if you 'collapse' a dodecahedron there is one magical spot where the edges are all the same length, and they can be positioned such that all of the new, jagged faces are completely flat. Julian shared this interesting new shape with mathematician Bill Gosper and they both assumed it was a rediscovery of an already known shape. But, try as they might, there was no evidence anyone had made this dodecahedron before! They had been calling it the 'tympanohedron' because 'tympanum' is Latin for 'drum' and it looks a bit like a drum. However, it was a bit of a mouthful and so I started calling it the 'dodecahedrum'.

I play a modest role in this story because Bill Gosper was having a hard time getting the dodecahedrum included in the Wikipedia page for dodecahedrons. Rightfully so, since Wikipedia is not a place to publish new research and the dodecahedrum did not appear anywhere else online. It fell between two stools: it was significant enough that people wanted to hear about it, but not quite novel enough to warrant a research publication (oh, and the discoverer was no

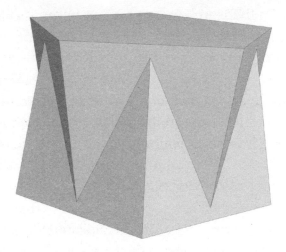

The dodecahedrum was an instant hit.

longer fifteen and had moved on with his life). Not for the first time, it was up to YouTube to break the maths news.

To make the video I asked a crafty friend to build me a 3D model of the dodecahedrum. Then, on a whim, it occurred to me we should make a model of the shape which also doubles as a drum. In the end I commissioned three different makers to build drums in the shape of the dodecahedrum. I then asked a drummer to play them all and give a drum review to go with my geometric review. The results were pretty clear: great shape, bad drum.

The dodecahedrum was surprising because 'a pentagon dodecahedron with all edges the same length' is such a low-hanging combination of constraints it was amazing no one else had stumbled across it. It even looks a bit like the endo-dodecahedron, and can be easily made by accidentally sitting on a regular dodecahedron. But, as far as we can tell, no one ever thought about such a shape before 2011 and no one decided to make a physical model of one until I did in 2023. Julian's equilateral shape was also an enjoyable shape to behold (less so to play).

The moral of a sort is, if you are prepared to investigate some interesting constraints, there are plenty of shapes out there waiting to be discovered. It's actually so common to find a new shape that having one take off in the media, like the scutoid or the Hat tile, is the exception rather than the rule. For example, in 2018 a maths teacher in the US named Robert Austin was curious about whether there were any shapes made only from kites and rhombuses. We've met rhombuses before – skewed squares with four sides all the same length – and kites are also quadrilaterals but with adjacent pairs of sides the same length. They're shaped like, well, a kite.

Robert fired up some geometry software called Stella 4d: Polyhedron Navigator and messed around for a while. He found eight 'kite-rhombus solids' and made a blog post about the shapes on his website. And that was it.

These 'new' shapes sat largely unnoticed on Robert's website until 2022, when I was looking around for some shapes with an unusual but specific collection of qualities. There is a festival in the Cotswolds in England called The Big Feastival, which is held every year on the farm of Blur bass-guitarist Alex James. The Big Feastival is already a celebration of Alex's two passions of food and music, but within it there is

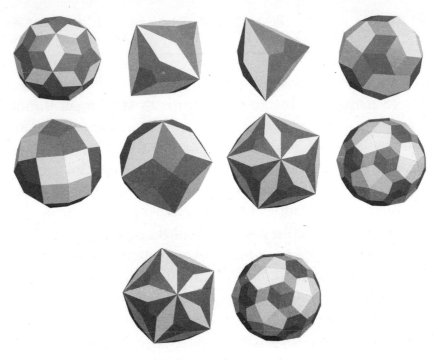

Hey, you got kites in my rhombuses!

a combination bar and dance-space called the Cheese Hub which narrows in even further, combining Alex's niche interests in cheese and dance music. But it was still missing his third love: mathematics. Alex has always been into maths and science; the call signal for the Beagle 2 Mars lander was a tune written specifically for the occasion by Blur.

Alex asked me if I could dream up a maths installation to complement DJs and cheese. For me the solution was obvious: a geometric disco ball. I get why traditional disco balls are just a bunch of square mirrors stuck to a sphere, but there are so many better ways to achieve the same effect. So I set about looking for a shape which was close to a sphere, but didn't have so many faces that we couldn't build and install a few during the festival (I wanted the festival-goers to help build the balls). I also wanted it to look 'mathsy' to the lay public. I didn't want people to just think these were unorthodox disco balls; I wanted them to easily see there was something mathematical going on.

Finally, I wanted the shape to have chirality. Which is to say, coming in left and right versions, like our hands. If you look at your left hand in a mirror it will no longer look like your left hand; it is now flipped and looks like your right hand. Hands are different but symmetric. Many shapes do not have this handedness: if you look at a regular dodecahedron in a mirror it looks like . . . a regular dodecahedron. Its mirror image is identical to the original shape.

My search happened to take me to Robert's website. I was not specifically looking for kite-rhombus shapes, but rather flicking through endless pages of image search results looking for anything with the right aesthetic flair. And the eighth of his eight solids caught my eye.

When Robert was exploring kite-rhombus options he decided to craft them by combining Archimedean solids with their dual Catalan shapes. He would nest each Archimedean solid inside its matching Catalan shape and then make a new shape by joining all of the exposed vertices together. Of the thirteen options, eight produced shapes made of kites and rhombuses (the other five were rhombus-only).

I say thirteen Archimedean solids: Robert actually checked all fifteen. I didn't have the heart to say before, but the snub cube and snub dodecahedron come in two flavours (which means we have even more shapes to exclude if we wish to keep to Archimedes's total count of thirteen). They are both made by expanding the cube/dodecahedron and then filling in the gaps with equilateral triangles. This process of 'snubbing' requires giving the square/pentagon faces a twist, and that rotation can be clockwise or anticlockwise. The results are mirror-image pairs and officially they are counted together as a single solid. But technically they are different shapes. They have chirality!

Robert's kite-rhombus process actually gives the same result for both versions of the snub cube; the chirality doesn't survive the transformation. For the snub dodecahedron it does! The result is a 150-face solid with chirality. It had enough faces to be a respectable disco ball and I could make pairs of these mirror balls. My plan was to hang them next to each other, rotating in opposite directions as if you were looking at one in a mirror. Which made them mirror mirror balls! To this day I am still disproportionately proud of the concept of making mirror balls in a mirrored arrangement.

When a snub dodecahedron and a pentagonal hexecontahedron get on really well.

Pair of balls but they are not identical! Which is perfectly normal.

Simon Pegg DJing under my balls.

For me, this is the final stage in the birthing sequence of a new shape. It starts when someone explores some interesting constraints or a novel application, and ends when someone actually makes a model of the shape. I know that making a physical model doesn't make the shape any 'more real' in a technical sense, but I think it does in a human sense. As I watched the mirror mirror balls spin above the crowded dance floor, it was a special moment. Alex James said 'this may be my proudest achievement', which I suspect is an exaggeration but I'm going to take him at kite-rhombus-face value.

Computer Shapes

If there is one theme to recently discovered shapes, it's the use of computers. Obviously a computer is not required to discover shapes – Plato, Archimedes and Johnson all did perfectly well without one – but it sure does speed things up.

In 1965 the Swiss mathematician Jean-Pierre Sydler needed a 3D shape made completely from 90° right angles, except for a single 45° angle. It was an important link in a proof about what shapes can be made from the parts acquired when a cube is sliced up. We're not going to dive into that actual proof but I can say that, after a bunch of analogue working out, Sydler was able to describe a way that a 'Sydler solid' could be made with the required angles. It looks a bit like an Escher nightmare, but every one of the angles you can see in it are 90° apart from a single 45° wedge.

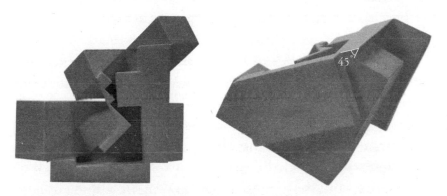

The beautiful mess that is the Sydler solid. That pointy wedge ledge on the left is the 45° angle.

In 2022 mathematician Robin Houston came across a 3D model of the Sydler solid and decided there must be a better version. Not that one was needed: the 1965 monstrosity proved that such a shape was possible, which is all the greater proof required. Nobody bothered to continue the search.

Until Robin fired up his computer and almost instantly found a much nicer one.

Robin's shape with a 45° angle in that overhang bit at the top.

This is the extent to which computers are a game changer in the field of shape finding. It still takes the same mathematical insight to work out the constraints for a hypothetical new shape and how the search could be conducted, but those steps can then be done much faster on a computer than with pencil and paper.

A perfect example is the hunt for the biggest shape. Not the biggest unbounded shape – that would take us back to space orbs – but the biggest shape that can fit within a sphere for a certain number of vertices. If the idea of trapping the biggest polyhedron in a small sphere upsets you, bear in mind that you can imagine an arbitrarily large sphere and no one can stop you.

We're skipping right over the 2D case because it is boring. What is the biggest shape with five corners you can fit in a circle? A regular pentagon. Six points? A regular hexagon. n points? A regular n-agon. It's always just the regular shape with that many corners. Evenly spreading the corners around the circle will always give you the maximum area. In 3D it gets far more complicated.

For a start, there is no easy way to equally space any given number of points on a sphere. This is named the 'Thomson problem' after physicist J. J. Thomson, who was originally interested in arranging electrons within an atom. The physics was incorrect (electrons do not hang out on a sphere) but the maths ended up being enduringly interesting. The solutions are easy for the cases of 4, 6, 8, 12 and 20 because they match the number of vertices on a Platonic solid, which provides a perfectly spaced arrangement. Leaving us with only the infinitely many other cases. To this day, the Thomson problem has not been solved. I'm sure golf-ball manufacturers are watching with keen interest.

Even if one day a mathematician does solve the Thomson problem, that will not instantly solve the biggest-shape problem. It is possible for equally spaced vertices not to correspond to the biggest volume possible for a shape. The corners of a cube will equally space eight points on a sphere. But the cube is not the biggest volume possible!

If not a cube, then what is the biggest shape with eight corners? The shape made by two hexagonal pyramids back-to-back is a good start with a bigger volume. A cube-in-a-sphere has a volume of about 1.5396 (compared to the sphere's radius = 1) whereas the friendly pyramids have a volume of 1.732 (which is actually the square root of 3, see if you can work that one out!). That is almost 12.5 per cent bigger! And once we've proven that the cube is not optimal it opens up the possibility for shapes even better than the hexagonal pyramids.

Back in the early 1960s, a Stanford University master's student in computational analysis (what we would now call computer science) named Donald W. Grace figured out he could use computers to solve exactly this problem. Unlike

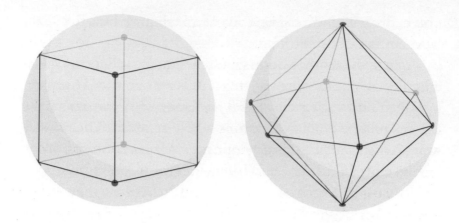

Robin looking for 45° shapes, Don didn't have his own computer he could just fire up. Stanford had only just received its new, expensive, room-filling Burroughs 220 computer in 1960 and it was hard to get time to use it, especially if you were just a master's student with dreams of hunting a shape. Don had to volunteer to take the night shift and run his code when no one else was around.

His plan was to give a computer the task of moving eight points around on a sphere. Don would start with eight dots in arbitrary locations and, at each step, the computer would calculate the volume enclosed by the points. To decide what the next step should be, the code would also calculate all the volumes of the possible new shapes if each point moved slightly in every possible direction. Which is a lot of potential shapes!

This allowed the computer to work out the 'gradient' of how the volume would change if any of the points were moved ever so slightly. The points would then be moved in the direction of the biggest new volume (which Don imagined as 'up the steepest gradient') and the process repeated. By looping through this over and over until no further minor tweaks could increase the volume, a maximum polyhedron would gradually evolve. It required more calculations than a human could do in a lifetime, but for these new-fangled electronic computers it would only be a bit of effort.

I don't know what happened when Donald Grace fired up the Burroughs 220 and ran his code. I imagine punch cards were flying around and magnetic tape wheels were spinning. A burst of computer activity in the dead of night on an otherwise silent Stanford campus. But whatever computing commotion there was, it eventually came to halt. Out popped a new shape. One never seen or imagined by humans before. A polyhedron with eight vertices that had more volume than any other eight-vertex shape previously known.

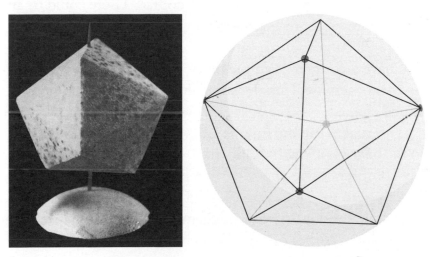

The model Grace used to visualize what the shape looked like. Pleasingly for us, it is made completely from triangles.

He published his findings in 1962 under the title 'Search for Largest Polyhedra'. In the paper, he outlines the mathematics he was using and then also acknowledges that the computer made things much faster.

-0.104528

Convergence was accelerated by letting the computer search for the multiplier, M, which caused the greatest increase in V [volume] at each iteration.

I read a lot of maths research papers from all different eras and, as I looked through Donald's paper, it was strange to see the sentence, 'The search was carried out using gradient methods on the Burroughs 220 computer.' It felt completely anachronistic because maths papers from this era never mention computers. Stanford didn't even have a computer science department until 1965. This was way ahead of its time.

In fact, I think that Donald's shape is the first shape ever discovered by computer.

There is a chance I'm wrong, of course. There could be an even more obscure geometric shape which was published before 28 August 1962, but I absolutely cannot find one. If you have a counter-example please do get in touch but, until proven otherwise, I am declaring this the first ever computer-found shape.

This makes the Burroughs 220 a very important computer in the history of maths. One that belongs in a museum! Sadly my comprehensive check of computer museums turned up exactly zero extant examples. I did a bit more digging, and it seems only about fifty-five of them were ever produced. The Burroughs 220 was one of the last of the vacuum-tube computers, which were soon superseded by transistor machines. Stanford received theirs in 1960, Don used it in 1962 and by 1963 it had already been replaced. No one knows where it went.

No one knows where any of them went. Except one. Well, parts of one. Sometime around 1970 a college somewhere in Illinois (we don't know which) was throwing out their now completely obsolete Burroughs 220. An employee at that college hated to see it scrapped, so they somehow dragged parts of it home and stored them in their basement for the next fifty years, after which it was put up for sale on eBay and bought by the owner of a film studio in Nebraska.

You see, even though the Burroughs 220 was a short-lived computer at the end of the vacuum-tube era, my goodness did it look like a computer. It really looked like a computer. If you're thinking of a 1960s-style computer, you are basically thinking of the Burroughs 220. So when computing departments stopped using them, as well as one of them making their way into a basement in Illinois, a bunch more made the move to Hollywood for careers as film props. You can see Burroughs 220 control panels in 1960s TV shows such as *Voyage to the Bottom of the Sea* and *The Land of the Giants*. One even held out long enough out to get a starring role in a 1980 episode of *Laverne & Shirley*.

But then sexier, younger computers came along and Hollywood discarded their Burroughs 220 parts, much as the computer departments had a few decades earlier. Until there was a second wave of film enthusiasts who wanted to recreate the look of classic mid-century sci-fi. Once such person was Bill Hedges of Cosmic Films Studio who spotted the eBay listing for the Burroughs and recognized it not for its significance to shape-finding history, but to TV and film history. He bought it and set it up in his film studio, which means the only extant parts of a Burroughs 220 anywhere in the world are in a film studio in Nebraska. I packed my bags.

I cannot imagine any other reason why I would visit the town of Lyons, Nebraska (population 824) but there I was. Visiting Cosmic Films Studio. In his retirement Bill had spent a lot of energy collecting and building classic TV sci-fi props, and had converted the Lyons cinema (which closed in 1985) into a worthy and functional home for his collection. He very kindly gave me a tour of the studio and in a room out the back, dressed to look like an underground lair, were the only four pieces of a Burroughs 220 in existence: a control console, two tape drives and the cupboard-sized tape-drive controller. Many key components are sadly missing, including the actual processor, but it's the best we have for now. Even though it cannot run, Bill had wired up the console so that at least the lights could turn on, making it look like the computer had sprung to life.

Main console of the Burroughs 220. The blinking lights are the closest thing this computer had to a monitor.

It might have been the exhaustion of the long pilgrimage, but this was yet another emotional moment. In front of me were the only known components from the computer model which found the first ever shape discovered by machine. I placed my offering on the Burroughs 220: a 3D-printed copy of Donald's shape to now live with the computer in its new home. In 1962 Donald had no idea he was the first in a long

Two tape drives and the cupboard of electronics required to run them.
Kind of like a room-sized thumbdrive.

line of people who would discover all manner of exciting shapes, using computers to render the searches achievable. Sixty years later, I feel we're still at the beginning of the shape-discovery journey.

Seven

GETTING TRIGGY WITH IT

T he Sphere in Las Vegas is a very impressive venue. A 366-foot-tall sphere, it is the most expensive building in the city to date. And if you read the Sphere's website, it has some unexpected praise for angles and something called 'the law of sines'.

> Like any global entertainment icon, Sphere would be nowhere without knowing our angles. The Law of Sines was used to calculate architectural angles across the building, from the pitch of the Atrium escalators to the curve of the archways in front of you.
>
> — Sphere Entertainment Co.

This is from the 'Science' section of the website, which also highlights that the Sphere was made as a triangle mesh ('What goes into constructing the world's largest Spherical building? A lot of triangles.') and even praises the finite element analysis used to build it, which we saw a few chapters ago. But the phrase 'law of sines' caught my eye because it casually makes the jump from angles to sine, a trigonometric function.

-0.190809

This book has been gradually building up to the moment when trigonometry arrives. Trigonometry is the turbo-charged version of geometry and has a reputation for being opaque and confusing. But at its heart it's as easy as riding a bike.

Like many middle-aged people with a growing realization that life is finite, a few years ago I took up cycling. It's a great hobby which provides all sorts of health benefits the vast majority of the time (with all the unhealthiness condensed into occasional, terrifying bursts). The part of the UK where I live is famous for its cycling, though not in a fun Dutch way, with loads of gentle, flat roads to roll along, but instead with spectacular hills designed to test how much matter your mind is capable of being over. These are the same hills which deterred the Romans and their straight-road ambitions.

One day I decided to go for a ride through an area called Hurt Wood. I should have taken the hint that this wood hurt. I soon found myself trying to ride up (the direction is import- ant) Barhatch Lane without realizing that it has been ranked as the second-toughest cycling climb in all of the Surrey Hills. My first hint was a street sign which provided a wel- come mental distraction, along with an ominous warning of what was to come.

The sign had a simple picture of a triangle and a percentage: 21%. 'That is oddly specific,' I thought, between gasps of air. Somewhere a road surveyor had refused to round to 20 per cent and I respect that. The sign is a warning that for every 100 metres you move forward along this road, you also go up 21 per cent of that distance. You're moving up a ratio of 0.21 to 1.

My next thought was, 'Huh, this road has a tangent of 0.21,' closely followed by, 'I wonder what that is in degrees?' And that is trigonometry. The whole subject in a single, exhausted thought.

That is not a good sign.

Trigonometry is all about ratios – in this case the distance up to the distance across – and how those values are just another way to measure the size of an angle. You can measure an angle in degrees, radians, fractions of a circle, grads (if you want to go absolutely grad) and now a new one: the tangent. This ratio is called the 'tangent' ('tan' to its friends) because in mathematics 'tangent' is often used synonymously with 'gradient' or 'slope'.

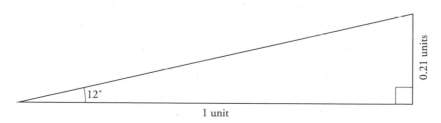

To work out what the angle of ascent is on a 21 per cent slope, I could draw any triangle where the rise is 21 per cent of the run and measure that angle. But that feels messy and imprecise. Instead I can just look it up. Which feels archaic and imprecise. But that is how it works.

−0.224951

In fact, I just looked it up two different ways. Out of respect for tradition, first of all I looked it up in a book. I grabbed my copy of *Chambers's Shorter Six-Figure Mathematical Tables* and found the tan table. I skimmed down to locate 0.21 and read off an angle of 11° and 51 minutes. Urg, of course it's in minutes, this book was first published in 1844 (I'm holding a relatively modern 1959 edition). But it could be even older: the earliest known trig tables are from around 1800 BCE. I did the quick conversion to get $11 + 51/60 = 11.85°$.

For a long time these trig tables were the only way to get trig values in a timely fashion. But they are imprecise because they can only list a handful of digits for each value. Then, at last, computers came along. So for completeness I used the 'tan' button on my cutting-edge smartphone's calculator, which told me that the angle is 11.85977912°. I know the sign designer liked their precision, but I suspect that is too much detail, even for them. Let's just say I was cycling up a triangle with an angle of around 12°. The important point here is that a ratio of 0.21 and an measurement of 12° are two ways of representing the same angle.

The creators of the sign could have put 12° on the sign and called it a day. But instead they went for 21 per cent. I assume they decided that was a more intuitive way to express steepness. Which I can kind of agree with: humans are pretty bad at guessing what angle ground they are on. Most people would not blink if you said a road was on a slope of 20°. It sounds like a small angle, but that would actually be a dangerously steep road! 12° is enough for a warning sign and the steepest roads in the UK max out at about 18°.

If the worst roads in the UK are 18° (a 32.5 per cent slope), and driving on a slope like that can be dangerous, then we have a problem when 18° doesn't sound like a scary number. A sign claiming there is a 30 per cent incline definitely sounds a bit more terrifying. So I can see why the tangent has been picked as the angle units for street signs. It's not the only way the UK does it, though. If you look up the official traffic signs in the Highway Code (as published by my good friends at the Department for Transport) you will see that, as well as percentages, 'Gradients may be shown as a ratio i.e. 20% = 1:5'.

Gradient signs on UK roads used to use ratios like 1:5 or 1:10, which indicated that for a rise of 1 you need to go forward 5 or 10, but these were phased out in the 1970s. Partly because it takes longer to read and understand something like '1:5' compared to a snappy '20%'. More importantly, in this system smaller numbers mean a bigger slope. If a car drives from a 1:10 section of road to a 1:5, the incline is actually twice as steep, despite the number getting smaller.

Even though the transition to the modern signs started about half a century ago, I've heard there are still some old-style hold out signs. And since the Department for Transport still feels the need to include a footnote about non-standard gradient signs (the only type of sign to get that kind of disclaimer in their official documentation), I feel like that rumour is pretty solid.

Sadly I am yet to find an extant '1:n' gradient sign and see it with my own eyes. A problem I only expect to have for the first edition of this book; as soon as it is published people will email in and let me know where such a sign exists. Hence, I'll leave a spot blank for the inevitable photo of me smiling and pointing at such a sign.

There is one last thing about the tangent signs I find pretty entertaining: people guessing what a 100 per cent sign would

This was the most uplifting moment of my life.

represent. If you take a look at the earlier triangle and think it through you can deduce that 100 per cent gradient is an angle of 45°. But when presented out of the blue people have all sorts of responses, including thinking it represents a vertical wall. Like some kind of Wile E. Coyote road terminating in the side of a cliff.

During a maths event for high-school students, we once gave this as a multiple-choice quiz question. Of the 590 responses we received, 51 per cent of them said an incline of 100 per cent would represent a vertical wall. When, many years ago, an annoyed *Guardian* reader in Glasgow wrote in to complain about these new-fangled gradient street signs, claiming the percentages were meaningless and the old '1:n' system was superior, it sparked a raging argument in the comments section. This resulted in one of my favourite ever confidently wrong comments on the internet, when someone was pushed on what a 100 per cent gradient represents: 'with 100% being a brick wall which it seems most of you have hit'.

Sine Language

Looking back at my diagram of the 12° road, you'll note that we used the ratio between the 'rise' (the opposite side to the angle) and the 'run' (the bottom side, adjacent to the angle) to get our tangent ratio. But there are two perfectly good ratios we've left on the table. If we know the length of the long 'hypotenuse' side as well, which a quick bit of Pythagorean Theorem reveals is 1.0218 (a surprisingly small increase on the base!), we can now find the ratios of the opposite and adjacent sides to the hypotenuse. And those, my friends, are the trig ratios of sine and cosine.

You may know them better by their nicknames: sin and cos. For some reason 'sin' keeps the same pronunciation as the longer 'sine' (I assume to make jokes like 'living in sin' just a little bit less funny). Whereas 'cos' does not sound like the start of cosine, but rather like a shortened version of 'because' (I assume to make the joke 'just be cos' slightly more funny). Fulfilling the conservation of funny between sine and cosine.

Actually a lot of things are conserved between sin and cos, because they are two sides of the same triangular coin. The names of the sides of a right-angle triangle have now crystallized into 'hypotenuse' for the longest side opposite the right angle, and 'opposite' and 'adjacent' for the other two sides relative to either of the other angles. But because those two angles are themselves linked (they add to 90°) the whole situation can feel a bit circular, the sine of one angle being the cosine of the other angle.

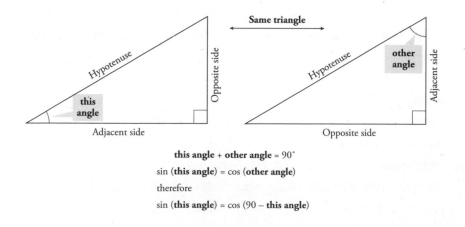

this angle + other angle = 90°

sin (**this angle**) = cos (**other angle**)

therefore

sin (**this angle**) = cos (90 − **this angle**)

By now you may have experienced one of the most common barriers to the world of trigonometry: remembering which name represents which ratio. This could have been solved if instead of 'sine' we'd called it 'oppositedividedbyhypotenuse', but it is too late for that. Hence generations of school students have learned SOHCAHTOA as an acronym for Sine: Opposite / Hypotenuse, Cosine: Adjacent / Hypotenuse, Tangent: Opposite / Adjacent. Often said as a single word, 'sohcahtoa', it is not a bad way to blindly remember which ratio is which. Or you can come up with: some other helpful, catchy acronym heuristic, that often assists.

$$\sin = \frac{\text{opposite}}{\text{hypotenuse}} \qquad \cos = \frac{\text{adjacent}}{\text{hypotenuse}} \qquad \tan = \frac{\text{opposite}}{\text{adjacent}}$$

If all of that caused you to glaze over, or have flashbacks to a teacher yelling 'SOHCAHTOA!' at you, do not panic. There is no need to memorize which ratio has which name (unless you're about to sit an exam). People doing actual mathematics can just look them up as required, and if you do memorize them it will be accidentally through frequent use.

Memorizing all the ratios is also a bit futile because there are so many of them, way more than three. There is a different name for each ratio when they are flipped upside down (each with its own three-letter abbreviation). Secant ('sec') is hypotenuse divided by adjacent, cosecant ('csc') is hypotenuse divided by opposite and cotangent ('cot') is adjacent divided by opposite. But you don't see kids being forced to memorize SHACHOCAO, even though it's a lot more shachy.

Arguably you don't need names for these extra ratios as they are just the inverse of sin, cos and tan. In that regard, they are like a host of other outdated trigonometric byproducts:

exsecant, excosecant, coversine and, my absolute favourite, haversine. These can all be calculated from the original trio. The haversine of an angle is equal to squaring the sine of half that angle. These functions are relics of the pre-computer days of looking up trig functions in a large, printed table. To save extra tedious calculations, these results of common trig calculations got their own names and were listed separately.

To pile on the stack of different functions, we also have the wealth of 'trig identities'. The trig ratios are all so interlinked, a bit of algebra can reveal all sorts of relationships. For example, you can get the value of tan by dividing the sin by the cos of the same angle. Combine with the Pythagorean Theorem and you'll find that if you add sin squared to cos squared (for the same angle), the result is always one. There is a whole world of other trig identities like this, which are clever ways to link sin, cos, tan and friends together. Including identities for splitting one angle into two or three smaller ones. Here's a small sample of them:

Classic

$$\tan(\theta) = \frac{\sin(\theta)}{\cos(\theta)} \qquad \sin(\theta)^2 + \cos(\theta)^2 = 1$$

Antiquated

$$\text{haversine}(\theta) = \sin\left(\frac{\theta}{2}\right)^2$$

$$\text{exsecant}(\theta) = \sec(\theta) - 1$$

Multi-angle

$$\sin(A + B) = \sin(A)\cdot\cos(B) + \cos(A)\cdot\sin(B)$$

$$\tan(A + B + C) = \frac{\tan(A) + \tan(B) + \tan(C) - \tan(A)\cdot\tan(B)\cdot\tan(C)}{1 - \tan(A)\cdot\tan(B) - \tan(B)\cdot\tan(C) - \tan(C)\cdot\tan(A)}$$

-0.325568

The classic identities are handy if you want to swap from one trig function to another. Multi-angle ones, such as that tan behemoth above, can be useful if you have an angle like 34° which you'd like to split into three smaller angles of, say, 13°, 1° and 20°. Which is not going to happen often enough that it's worth memorizing. The goal for students of trigonometry is to gain a sense of the flavours of these types of trig identities. Then they can search for them in the chocolate box of options when they are later trying to do some complicated geometric working-out.

It is a shame that so many students get put off trigonometry because of the seemingly pointless memorizing of these strange ratios. In reality these trig functions, and the host of relationships which link them, are new tools we can throw into the rule-box along with our previous triangle laws. Trigonometry means that we can solve the missing parts of a triangle even more easily, and in more situations. Trig is why so many problems can be cracked by reducing them to triangles.

In the introduction of this book I mentioned the oilfield worker who had to learn geometry to be promoted to the role of driller. What I didn't mention is that the next promotion, up to directional driller, required trigonometry. One day a superior came to them and said, 'Ever work with trig?' They had to sit down and master trig to advance their career, and wished they had learned it earlier. Trigonometry really is a massive upgrade on geometry.

In fact, it has been a bit of a personal challenge to not bring up the trig ratios of sin, cos and tan until now, because they feature in so many triangle situations. The equation for the height of the balloon above the pigs used the tan of the viewing angle; the internal friction inside the Dimorphos asteroid was calculated using the sin of the angle of repose; the amount gravity was trying to tip my MotoGP bike over is the cosine of the angle it was on; Paul's UFO calculations used both sine and cosine; William Thomson's investigation into the truncated octahedron filling space required the square of tan; and the proof of Grace's 'biggest polyhedron' shape used the cosine of its internal angles.

That is just one example from every chapter so far and I could have listed many more. Trigonometry is so powerful, you'll struggle to find a modern triangle calculation which doesn't involve a trig ratio.

Won't You Take Me To, Function Town?

All those examples from the previous chapters are great and all, but forget right-angle triangles: I want to focus on the idea that trig ratios represent an interesting new way to represent angles. Every trig function can take an angle and convert it into an equivalent value. Here's a table of a few functions and the values they spit out between 0° and 90°.

DEGREES	SINE	COSINE	TANGENT	HAVERSINE
0°	0	1	0	0
10°	0.1736 . . .	0.9848 . . .	0.1763 . . .	0.0075 . . .
20°	0.3420 . . .	0.9396 . . .	0.3639 . . .	0.0301 . . .
30°	0.5	0.8660 . . .	0.5773 . . .	0.0669 . . .
40°	0.6427 . . .	0.7660 . . .	0.8390 . . .	0.1169 . . .
50°	0.7660 . . .	0.6427 . . .	1.1917 . . .	0.1786 . . .
60°	0.8660 . . .	0.5	1.7320 . . .	0.25
70°	0.9396 . . .	0.3420 . . .	2.7474 . . .	0.3289 . . .
80°	0.9848 . . .	0.1736 . . .	5.6712 . . .	0.4131 . . .
90°	1	0	∞	0.5

These functions would be functionally useless if all they did was produce another linear way of measuring angles. Degrees are linear: if an angle gets twice as big, its value in degrees just doubles. How droll. These trig functions are all over the place, which makes them a terrible way to measure how big an angle is, but they reveal other properties of an angle. Check out this smörgåsbord of ways trigonometric functions save the day:

1. Components

Sine and cosine are the 'anti-Pythagoras'. If you have a right-angle triangle, it is possible to find the length of the hypotenuse by squaring each of the two short sides, adding them together and then taking the square root. But that process cannot be reversed without introducing some ambiguity. There are multiple possible combinations of different-length short sides which all have the same hypotenuse. However, if you know one of the angles in the triangle, you can use sine and cosine to accurately reverse the calculation and find out how long the original sides were.

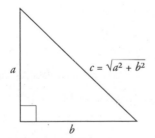

$$c = \sqrt{a^2 + b^2}$$

$$a = c \cdot \sin(\theta)$$

$$b = c \cdot \cos(\theta)$$

Finding the long side 'c' using 'a' and 'b'.   Finding the short sides 'a' and 'b' using 'c' and one angle.

The sine of an angle gives the ratio of the opposite side to the hypotenuse, and cosine gives the ratio with the adjacent side. This is arguably just the definition of those trig ratios, but it is still one of their most useful applications because of how often we want to convert to and from coordinates.

Thinking in terms of the basketball shot data from the NBA, I was using Pythagoras to take the x and y coordinates on the court and calculate the distance and direction from the basketball hoop. If instead I had started with the location of the shot relative to the hoop, I would have needed to use sine and cosine to work back to the coordinates. So much data in our everyday life is stored as coordinates that it is invaluable to be able to quickly flip from distance and direction to coordinates, and back.

Pixels in a digital image are stored as pairs of dimension coordinates: how far across the image and how far up. The location of something in space is often stored as three coordinates: how far across, how far back and how far up (akin to the position on a map's grid, plus the altitude). I was able to combine both of these things when I was programming the lights on my Christmas tree. I was annoyed that strings of decorative lights always illuminate in patterns that run along the direction of the wire joining them. I wanted to have lighting patterns which didn't follow the physical wiring, but rather the geometry of the tree. The geometree.

In recent years, you've been able to buy decorative lights which come with an app you can use to project patterns onto the surface of a Christmas tree. But where is the Christmas spirit in that? I wanted to do this myself, and I wanted to make sure I had full 3D coordinates so I could control the lights deep within the tree as easily as those on the tips of the branches.

My solution was to dump a chain of 500 LEDs onto my Christmas tree with no care for where the wire was going. I worried only about making sure every branch in the tree was completely festooned. I then plugged the lights into my laptop so I could use software to turn them on and off individually. One by one, the computer code I wrote would turn on an LED and use the webcam to take a photo of the tree in a dark room. The x and y coordinates of the brightest pixel in the photo would give me the horizontal and vertical location of that LED. Once I had completed all 500 LEDs, I physically rotated the tree 90° and repeated the process to get the other direction as well. At the end of the world's most boring Christmas-light display, I had the 3D coordinates of where every single LED was.

I could now program any crazy lighting patterns I fancied. An easy one was to take the vertical coordinate of every LED and turn them on and off according to their height. The idea was to have a wave of light travel up the tree. I wrote some code so that all the lights within a certain height range would turn on, and then moved that range up the tree. It was as if a flat plane was moving through the tree and turning on all the lights it touched.

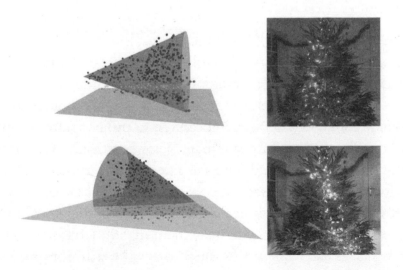

Visualization of the code moving a 'light plane' through the 3D coordinates of the lights to turn them on in waves, and what that looks like on the actual tree.

-0.406737

Not content with just waves of light going straight up, I had a think about how I could code waves to go in all sorts of random directions. Like all things in maths, there were multiple ways this could be solved, but I decided that my 'rising wave' was working so well I would not touch that code. Instead, I would move a virtual copy of the tree around. Between each passing wave, my code would rotate the lights around the trunk of the tree by some random angle (which I called 'α') and then tip the whole tree over by some second random angle (aka 'θ'). If I sent the rising flat plane over this tilted, virtual tree and activated the light on the physical, right-way-up tree, it would look like the waves of light were moving in random directions.

The issue was how to get the new vertical coordinates of the LEDs once the tree had been spun and tipped. It's a bit more complicated than using just a sine or cosine to get the 2D component, but I was able to code up this trigonometric beauty:

$$z_{new} = \sin(\theta)[x \cdot \sin(\alpha) + y \cdot \cos(\alpha)] + z \cdot \cos(\theta)$$

Do not worry if that equation makes no immediate logical sense to you. It doesn't to me, either. If you showed me that equation with no context, I absolutely could not tell you that it describes the vertical coordinate of a Christmas tree that has been rotated and tipped. Because I don't waste my life memorizing such things. I guarantee that anyone who can reflexively tell you what that does has spent so long working with such equations they have accidentally memorized them.

It's just that I knew that there would be a trigonometric equation to solve my problem. I looked up the equations for both the rotation and the tip, multiplied them together alge-braically* and extracted the vertical coordinate. The power of trigonometry is not memorizing equations, but rather knowing you can look them up and having faith they will solve all your problems. In cos we trust.

2. Beach Rescue

At the same high-school shows where we did the '100 per cent gradient' quiz we also sometimes give a geometry puzzle based on the idea of rescuing someone who is in the ocean. One person starts on a sandy beach and needs to reach a second person in the ocean as quickly as possible.

If both points were on land then the solution is easy: run directly between the two points. Once water is involved it gets complicated, but people can typically run faster on the beach than they can swim in the ocean, so it might be worth running a bit more on the beach to reduce the distance swum in the sea. We leave it to the students to guess or work out the optimal point to switch from sand to surf.

Spoiler: the solution is not to completely minimize the water distance – as that requires so much running it more than negates the time saved by reducing the swim – but somewhere in between. Depending on what maths the teenagers have previously learned, they typically write down some equations to compare all of the distances and then work out how to minimize the total time. Which does work.

But instead of thinking about this problem as the distance

* For you detail fans, I actually looked up the rotation matrices, multi-plied those together and took just the vertical component.

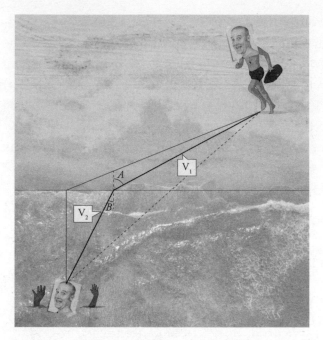

Angles a and b are exactly the same as if light moved as slow as your mass-filled mortal body.

run on the beach versus the distance swum in the ocean, it can be thought of as choosing the optimal angles to leave the beach and enter the sea. In a logical world, how much faster the person can run on the sand compared to swimming in the sea would determine the ratio of the angles a and b. If you look at the angles in degrees there is no obvious link. But if you switch from degrees to sine, the link is so simple it's embarrassing. The ratio is the same.

$$\frac{\sin(A)}{\sin(B)} = \frac{\text{speed on sand}}{\text{speed in water}}$$

The ratio of two speeds is exactly the same as the ratio of the sines of the two angles. The reason why is mildly complicated but the point is, if you can run twice as fast as you can swim, you should enter the water when the sine of the beach angle is twice the sine of the water angle. And this doesn't just apply to running on the beach.

It transpires that the optimal point to run into the ocean is exactly the same as the angle of refraction we saw back in Chapter One when looking at rainbows. If you consider the human as a photon moving at different speeds through different mediums, the shortest path is the one a photon would take to refract at the shoreline and hit the target. The exact same use of sine values gives us the refraction angles for a rainbow.

This relationship is often called Snell's Law, after Dutch mathematician and massive triangle fan Willebrord Snellius who wrote it down in the early 1600s (although it was almost certainly known before his time). It's a case where the shortest path to a solution, skipping the longer path of lots of working out, is merely to change all the angles into their sine values.

3. Of Trump and Trig

In August 2019 President Trump was shown a printout of an Iranian launchpad taken by a classified US spy satellite. This would be the kind of top-secret meeting that the Oval Office was built for; we civilians should have had no idea it even took place. But Trump was so impressed with the image that he whipped out his phone, took a photo of it and tweeted it. Depending on if @realDonaldTrump is banned from Twitter (or its later manifestations), you can sometimes see the original tweet in all its glory.

Accidentally released classified or sensitive information is like blood in the water for the internet, and a host of sleuths

Donald J. Trump ✔
@realDonaldTrump

The United States of America was not involved in the catastrophic accident during final launch preparations for the Safir SLV Launch at Semnan Launch Site One in Iran. I wish Iran best wishes and good luck in determining what happened at Site One.

1:44 pm · 30 Aug 2019 · Twitter for iPhone

descended on the picture in a calculation frenzy. Whatever spy satellite had taken this image was top secret, and armchair analysts realized they could reverse engineer the photo using angles and trigonometry to learn about the spacecraft responsible.

Once such angle-angler was Cees Bassa, an astronomer working in the Netherlands. He wanted to know exactly where the satellite would have been in the sky when it took the photo. The first step was to get a reference overhead-shot of the facility. It's not exactly a secret location and you can find it on Google Maps fairly easily (35.234618 north, 53.920943 east), albeit at low, civilian resolution. There are four towers around the pad, and a careful measurement reveals they are not exactly aligned with the compass directions but are 12° around in the clockwise direction. In

-0.484810

Trump's photo they appear to be 3.8° anticlockwise from the direction the camera was pointing. If we add 12° + 3.8° we now know this secret spy camera was on a bearing of 15.8° from north.

But how high was the camera in the sky? Again, the photo reveals all. In Trump's image the circular launchpad looks like an ellipse. Which makes sense: the satellite was not directly overhead, and if you look at a circle from the side it takes on a bit of an oval shape because of perspective. Cees Bassa worked out the exact shape of the oval-view in the photo and calculated it must have been taken from a camera on an elevation of 46.2°.

While the US military do not publish the details of where their spy satellites are, amateur astronomers keep a pretty close eye on the sky and do track the movements of anything up there. Bassa checked the logs and, sure enough, the satellite USA 224 passed right through a spot in that exact direction relative to the Iranian site at roughly the time when the photo was taken.

So a basic use of angles allowed civilians to confirm that this specific satellite flying around the Earth was a US military satellite. Which is very funny, and if that was the whole story you would be reading this in Chapter Two. But this is the trig chapter, and sine allows us to learn something about the internal workings of this clandestine American spacecraft. We can calculate how big its telescope mirror is.

For a start, the image is obviously in better resolution than normal satellite images. Thanks to USA regulations, no publicly available satellite imagery can have a resolution better than 30 centimetres per pixel (roughly 1 foot). All the detail from each 30-centimetre by 30-centimetre patch on the earth gets blurred together into a single pixel. Which is not the case in this image. What appear as blurry paths on Google Maps are now individually recognizable stairs. The shadowy towers are now resolved into individual struts. Going off this detail we are definitely seeing 10-centimetre-or-less per-pixel resolution. And that is merely the resolution the image was printed at! Potentially the digital image original was even more detailed, but all we have is the physical printout handed to Trump.

We can take that inferred resolution and calculate how big the camera on board must be using this one simple equation.

Size of camera = 1.220 × wavelength ÷ sin(*A*)

That is the whole equation. If you know the length of the wavelength of light (which we do), this gives a direct linear relationship between the sine of the smallest resolvable viewing angle, A, and the size of the mirror (the mirror size will be in whatever units were used for the wavelength). Don't stress about where that 1.220 ratio comes from, it's the combination of a bunch of complicated constants we do not need to worry about.

The spy image Trump tweeted showed detail down to a level of about 10 centimetres which means, at the altitude we know the satellite was orbiting, the camera lens must be around 2.5 metres wide. That is a strong enough telescope that you could stand in England and read a newspaper in France – though you'd need someone closer to turn the pages for you. It also confirms the theory that these spy satellites are flying around with exactly the same mirror set-up as the Hubble Telescope, which was launched with, you guessed it, a 2.4-metre mirror.

So it turns out that Trump was a bit of a national security

risk. Something you didn't need a 2.4-metre telescope to see coming.

New Rules

Baseball is famously a sport full of statistics. It's where the 'moneyball' revolution of using data analytics started, which has since spread to all other professional sports. But as much as I love stats, for me baseball is also a game of geometry. And for a sport with a simple 'diamond' shaped field, it is more complicated than you'd expect.

The geometric curveball is that a so-called baseball diamond is not actually a diamond. Mathematically a diamond is a four-sided shape where all sides are the same length, and is synonymous with 'rhombus'. In the case of baseball, the diamond is a square with sides of exactly 90 feet each. It's typically drawn rotated by 45° so that home base is at the very bottom, but I'm going to sketch it orientated to look like the square it's trying to be. The issue is that first and third base are placed completely inside the corners of the square but second base is centred on a corner. This means that the edges of the base are beyond the square, and so the 'diamond' is technically formed by the line which joins the far corner of first base with the outmost corner of second base.

I was curious about the effects of the second-base placement: the actual distance along that side of the 'diamond' must be longer and the angle at first base will be slightly bigger than the right angle it claims to be. So I drew a sketch of the 90-foot line between first base and second base, along with an exaggerated second base. The beauty of trigonometry is that diagrams no longer have to be to scale. I could draw a meticulously scaled diagram and try to measure it, but

-0.529919

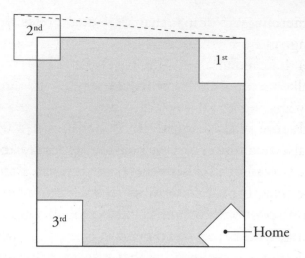

Not-to-scale diagram of how the bases are loaded.

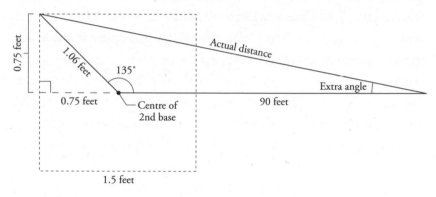

Exaggerated diagram of the top of the diamond, so it's easier to touch base.

instead I can draw a diagram where the fiddly bits are zoomed in to make it easier to see the geometry. I can then calculate the missing lengths and angles.

Looking at my triangle, we have two lengths and one angle. We know the main length to the centre of second base is 90 feet. Since 2023 the bases in Major League Baseball (MLB) have been squares with sides of 1.5 feet. Using the Pythagorean Theorem I can calculate the distance from the centre of the base to the outer corner as 1.06 feet. I know the angle is

135° because it is 45° more than 90° (or 45° less than 180° depending on how you like to look at it). I'm now having more fun than actually watching baseball.

I've talked a lot about 'solving' a triangle, meaning calculating any missing lengths and angles. This is exactly what we need to do in this case, to find the 'actual distance' and 'extra angle'. I also said that as long as half the values are known (in this case, two sides and one angle) we can calculate everything else. However, I've been real vague about how those calculations are done. This is because they require not only trigonometric functions but two new trig rules as well.

Trigonometry gives us access to even more triangle relationships in the same style as Pythagoras. These are not just trig identities like we saw before, which can be used to flip from one trigonometric function to another. These are relationships which always hold true in all triangles. Sine and cosine get one each.

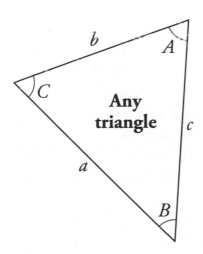

Cosine rule

$$c^2 = a^2 + b^2 - 2ab\cos(C)$$

Sine rule

$$\frac{\sin(A)}{a} = \frac{\sin(B)}{b} = \frac{\sin(C)}{c}$$

The cosine rule looks a lot like the Pythagorean Theorem but with an extra '$-2ab\cos(C)$' on the end. Because it is. The limitation of the Pythagorean Theorem is that it only applies to right-angle triangles. The cosine rule is an upgrade, allowing the same method works on all triangles. Whereas the Pythagorean Theorem works if you know two edge lengths on either side of a right angle, the cosine rule means if you know two sides flanking any angle you can calculate the other length. (And very neatly because $\cos(90°) = 0$; for the case of a right angle, the correction factor is zero and the cosine rule reverts back to the Pythagorean Theorem.)

The cosine rule allows us to calculate how long the actual distance on the side of a baseball diamond is. There are two sides of the triangle with lengths of 1.06 feet and 90 feet, and an angle of 135° between them. If those values are plugged into the cosine equation, the opposite side comes out as a length of 90.753 feet long. Which makes sense, it's slightly longer than the 90 feet to the centre of the base.

The other great thing about the cosine rule is that if you know all three sides of any triangle it can tell you what the internal angles are. This was one of the bits of maths the machinist I mentioned in the Introduction used as an example of the trigonometry they use all the time. A client wanted a part made with a very specific angle, and so the machinist deployed the cosine rule to measure three distances on the finished part and used trigonometry to prove the internal angle must be within the required tolerances.

The sine rule is the 'law of sines' that the Vegas Sphere was very excited about. It states that for any triangle the sine of an angle divided by the length of the opposite side is the same for all three corners. I think of it as every triangle having its own 'sine rule value'. For the baseball triangle we have 135° opposite 90.753 feet and so $\sin(135°) \div 90.753 = 0.007791544$. Any other combination of angle and opposite side in the same triangle will have a value of 0.007791544, so if we know the side opposite the 'extra angle' is 1.06 feet we can work back and get an angle of 0.4735°.

Putting it all together: on paper the corner of a baseball diamond at first base should be 90° and it is then 90 feet to the far side of second base. In reality it is a 90.4735° angle followed by a 90.753-foot side-length. If the two sides of the 'diamond' which touch home plate are actually 90 feet and the two sides around second base are both 90.753 feet, then a baseball diamond would be more accurately called a baseball kite.

MLB is no stranger to changing the geometry of the game. The bases had been 15-inch squares since 1877 until the increase to 18 inches (1.5 feet) in 2023. First and third bases used to be centred on the corners of the diamond and were repositioned to be inside the corners in 1887. Somehow, second base has retained its original position until now. For the sake of neat geometry, the next time MLB change the field I'd love to see either second base brought in as well, or first and third put back to their original positions. That would make everything fair and square.

Sine of the Times

Trigonometric functions are capable of all these incredible and disparate applications because they provide complicated, nuanced information about an angle. But there is one substantial downside to that complexity: they are real hard to compute. Trig tables were invented in the first place because calculating the values was a real pain. Books of trig tables hung around for so long because even computers struggle with it. It took a long time before calculators were in any position to provide trig values, and even longer before they could fit in a pocket.

The only way to get a trig value is to iterate your way closer and closer until you're happy with the level of precision you are working with. If you have some angle, 'A', and you want to know what sin(A) is, you just need to do as many terms in this series as you can be bothered doing.* You alternate between adding and subtracting and those powers go up through all the odd numbers. The exclamation marks are factorials; they do not represent the endless excitement of computing infinitely many fractions just to get one value of sine.

$$\sin(A) = A - \frac{A^3}{3!} + \frac{A^5}{5!} - \frac{A^7}{7!} + \frac{A^9}{9!} \ldots$$

That infinite series converges quite slowly on the value of sine we so desperately want. Plus, doing progressively bigger and bigger powers and factorials, only to divide them, is not a computationally efficient way to go about this. What we need are other sine algorithms which get to the value a bit quicker. The original computer trig algorithm, designed to do just that, was called CORDIC and emerged during the 1950s. CORDIC is an acronym of COordinate Rotation DIgital Computer and uses clever computer operations to effectively move a point around on a circle until the x and y coordinates match the sine and cosine of the requested angle. It's effectively a very rapid game of higher-or-lower,

* For this equation to work as-is, A needs to be measured in units of radians and not degrees. I picked the radian version as it is way neater, but the same idea applies for any units.

and forty consecutive guesses are enough to get any sine or cosine value to ten decimal places.

The first calculator to provide trig values used the CORDIC method. It was Hewlett-Packard's HP-9100A desktop calculator, and it was a massive breakthrough because it wasn't massive and could sit on a desk without breaking through. If there was a theme park dedicated to the history of calculators (fingers crossed, one day . . .) then the HP-9100A would be the E-ticket ride with massive queues, people lining up around the block for a chance to take it for a spin. Although it was, ironically, created to avoid queues.

The prototype for the 9100A was designed and built at home by then-unemployed engineer Tom Osborne. He believed that a desktop computer was possible when many people didn't. Including his erstwhile employers, which was why they had recently fired him. Tom set up shop at home and carried on soldering his vision together. He designed and built his own circuits which he packed into a wooden box that he had painted with car paint (Cadillac Green Metallic). These 'lone hero' stories often gloss over the less glamorous, behind-the-scenes support, so I think it's important to note that Tom's wife Carol was earning the family's sole income during this time. She also later wrote the assembly-language program which coded the ROM chip in the HP-9100A.

Thanks to their joint effort the first working prototype burst into life on Christmas Eve 1964.

I remember the overwhelming realization that sitting in front of me on a red card table in the corner of our bedroom/workshop, sat more computing power per unit volume than had ever existed on this planet. I felt more like the discoverer of the object before me than its

-0.615661

creator. I thought of things to come. If I could do this alone in my tiny apartment, then there were some big changes in store for the world.

– Tom Osborne

After several fruitless meetings with various companies, an ex-colleague of Osborne lined up a chat between him and some of the high-ups at Hewlett-Packard, including Hewlett and Packard. They wanted to develop a calculator small enough to fit in the space on a business desk where a type-writer was traditionally housed, and Osborne's 'green machine' was about the right size. The director of the HP lab suggested the machine use CORDIC so it could also do trig calculations and the rest is calculator history.*

The HP-9100A hit the market in 1968, and a magazine advertisement for it in October that year is the earliest docu-mented use of the phrase 'personal computer'. This was the original PC. But, at $4,900, you'd better hope you worked at a company that could buy it for you (median US income was $7,700 in 1968, so it was the better part of a year's earnings). It was 'personal' insomuch as you could get it all to yourself. The early ads touted it as being able to 'relieve you of waiting to get on the big computer'. The 9100 was 'At your fingertips whenever you need it.' (Assuming your fingertips recently didn't need $4,900.)

Also right there in the ad is the boast that the 9100A is 'Willing to perform log and trig functions'. Top of the list of features. Those advertisers knew that nothing would cause units to fly off the shelf than people having trig functions at

* Fun computing history coincidence: the CORDIC equations for the HP-9100A were tested at Stanford on a Burroughs B5500 which was the upgrade to the original Burroughs 220 Donald Grace found his 'biggest polyhedron' shape on.

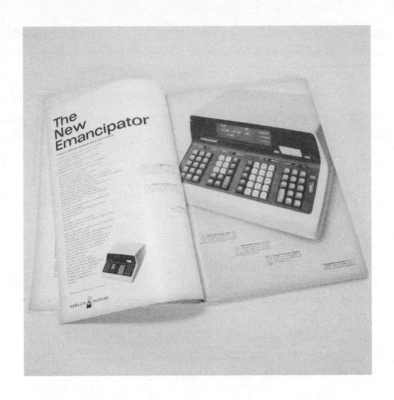

their fingertips. Shortly before the official release, HP took a pre-production 9100 to show the engineers at NASA's Jet Propulsion Laboratory. The marketing reps set up the 9100, pushed a button and it produced Bessel-function antenna patterns. The crowd literally went wild, with engineers jumping out of their seats and giving the calculator a standing ovation.

Easy access to computerized trigonometric functions was a game-changer in science and business. As computing time gets cheaper it can be used more frivolously on speculative projects. Which is where the fun happens. In the 1960s Donald Grace had to volunteer to take the night shift to get computing time for his shape-hunting adventures. The rise of desktop calculators made it far easier to try some calculations out of curiosity, even at NASA.

-0.642788

In the 1970s rocket scientist James Van Allen used an HP-9100A to dabble with some possible alternative trajectories for the already-launched spacecraft Pioneer 11. He found a new gravitational slingshot which would allow Pioneer 11 to whip around Jupiter in such a way as to hurtle in a Saturn direction. Pioneer 11 was redirected from across the solar system, and became the first human-made object to visit Saturn. The spacecraft Voyager 1 should have been the first visitor to Saturn but it was beaten out because a desktop calculator made doing trigonometric calculations easier.

Trig Tables are DOOMed

Computers have gone from strength to strength, but they are yet to be the death knell of the 4,000-year era of the trig table. In situations where computing power is limited, working out trig values from scratch is an unaffordable luxury and the pre-computer trig table once again comes to the rescue.

The computer game DOOM was released in 1993 and has since gone on to be considered one of the most important video games of all time. In it, you navigate a 'Doomguy' around a 3D world, taking on waves of enemy characters. It was the 3D-navigable reality (along with then-quite-realistic gameplay) which made it such a hit. It was designed to run on the most up-to-date computers available and a few years later it was also released for the newest range of 32-bit home consoles.

Then, in 1995, DOOM also came out on the Super Nintendo Entertainment System (SNES), which was a bit surprising. The SNES was designed for 2D graphics and was not powerful enough to render the 3D world of DOOM. This lack in power was partly compensated for by an extra processor in the game cartridge itself. But it was also down to a clever use of trig tables.

The developer Randy Linden released the original code for DOOM SNES in July 2020, which allowed retro game enthusiasts to comb through the programming to see exactly how the game was able to run. Baked into the code itself are five trig tables: sine, cosine, tangent, secant and reverse tangent. These were the trig functions required to convert between the 3D coordinates of the DOOM world and the 2D graphics of the SNES (in much the same way as I used trig functions to convert the 3D coordinates of my Christmas tree).

The trig tables were integrated into the game's code and took up 145 kilobytes of storage. That sounds laughably small now, but at the time it was a lot of drive space! The trig tables were the second-largest file in the entire game system. Which also gives you a sense of their importance. It was thanks to these clever trigonometry tables, a very ancient bit of technology, that the SNES version of this game was not doomed.

As I was writing about this DOOM code, I wanted to make sure I understood how the embedded tables worked but they were all in hexadecimal numbers which is great for computers but not so easy to read for us humans. I wrote my own method to convert the values into base-10 and when I checked them against the true values I noticed a discrepancy: a lot of the values had been rounded the wrong way. I got in touch with the creator, Randy, who confirmed both that I was the first person to contact him about the maths of DOOM and, yes, my suspicion was correct: the values had been truncated instead of rounded.

He linked me to the original computer code which generated the SNES DOOM trig tables and sure enough it was converting the answers from a 'double' type of number (which has decimal point values) to an integer by discarding all the fractional parts. But the difference in accuracy this produces is vanishingly small. As Randy said, 'The error isn't noticeable considering the resolution of the game . . . but, like all programs, there is room for improvement.' The 'bug' could be fixed with an extra line of code to round the intermediate value before turning it into an integer, or something similar. And it's not impossible: since the source code was released, someone has already made a patch to fix a bug which stopped the player from being able to rotate at the same time as strafing sideways. I admit it's not quite like getting the NBA shots fixed, but still!

Processors have come so far since the 1990s that it is now a common programming 'joke' to try and get DOOM running on hardware which shouldn't be able to run games. People have made versions of DOOM playable on thermostats, printers, the Zune and, of course, a TI-84 graphing calculator. I find it delightful that, while the original SNES hardware didn't have enough processing power to do the calculations in DOOM, DOOM can now run on a calculator.

Eight

WHERE ON EARTH?

Many years ago I was walking along the beach with my now-wife who wondered out loud how far away the horizon was. The beach is a good place to see the horizon; typically you have an uninterrupted view out over the water as it conforms to the Earth's spherical shape. And it is a good question: that most distant bit of water you can see – your own personal horizon* – how far away is it? The thought took my imagination as well so we quickly sketched out a diagram in the sand (not to scale) and drew ourselves standing on the Earth. Please enjoy my best reconstruction of what that sandy self-portrait looked like.

We designated the Earth's radius as a big 'R' and our puny human height as a small 'h'. Distance to the horizon was 'd' and, importantly, that line of sight will hit the surface of the Earth at a tangent, making a right angle with the Earth's radius. Some rearranging of the equations using the Pythagorean

---

* Horizons are a bit like rainbows in that everyone gets their own, depending on where they are standing to look into the distance. We'll assume, from here in, that my date and I were seeing exactly eye-to-eye.

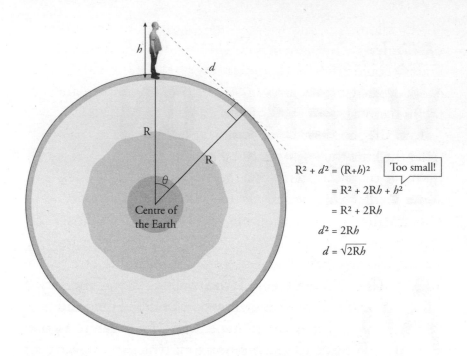

$$R^2 + d^2 = (R+h)^2 \quad \boxed{\text{Too small!}}$$
$$= R^2 + 2Rh + h^2$$
$$= R^2 + 2Rh$$
$$d^2 = 2Rh$$
$$d = \sqrt{2Rh}$$

Theorem got us close to working out what d was, but the algebra was getting messier than the resolution of sand – and our patience, as the Sun was setting – would allow. We decided to cheat and simplify things.

The 'h^2' bit represents our height squared, and it's the smallest value on that line. Both the R^2 and 2Rh terms involve the Earth's radius, R, which is huge compared to our heights. So, even though our height will make a difference to the final answer, we can probably remove the h^2 term from here because it is far smaller than everything else in the equation. In one motion, we brushed that section of sand clean and simplified the algebra.

The final step was to estimate the values of R and h so we could put them into the equation and get an estimate of d. We knew the Earth's diameter (aka 2R) is about 12,500 kilometres (because one metre was defined as one ten-millionth of the way from the North Pole to the equator, the circumference is

40,000 kilometres, and we worked back from that to estimate the diameter) and humans are close enough to 2 metres tall that we could use 0.002 kilometres for h. $12{,}500 \times 0.002 = 25$, and if we take the square root of that we get our final answer: the horizon is about 5 kilometres away.

In maths, as in love, it is important to know when it's best to let the little things go. In our case, it meant we could get an estimate of the distance to the horizon using a stick and some mental arithmetic. We did work out a more accurate answer later, and our estimate was happily quite close! My eyeballs are 1.7 metres off the ground (not 2 metres) and the Earth's diameter at that beach is 12,744 kilometres (not 12,500). Using cosine I can get the central angle in the Earth, 0.04234°, and the sine of that will allow me to calculate the distance to the horizon as 4.7 kilometres. Very close to our estimate of 5 kilometres. I hope that is enough to please all you closure fans. Oh, and we're married now. As a result of that and many other calculations.

The point is, humans have some natural urges, including the one of wondering how far away the horizon is. And wondering how big the Earth is. There are only two ways to answer those sorts of questions: a lot of walking or triangles. Even the triangles include a decent chunk of walking because, as we've seen before, you always need at least one side of a triangle.

Caroline Herschel was a powerhouse astronomer; in 1787 she became the first woman to have her results published in the Royal Society's main publication. Reading her memoir, I was struck by the list of mundane astronomical errands she had to perform: 'either to run to the clocks, write down a memorandum, fetch and carry instruments, or measure the ground with poles, &c'. No matter how lofty your objects of study, the ground always has to be measured.

The first probably accurate measurement of the Earth we know about was by Eratosthenes, a mathematician living in Alexandria, Egypt sometime in the third century BCE. As is so often the case, we have not found any surviving copies of his original working-out but from references in other, later writings we get the general story. Apparently Eratosthenes had heard that in the middle of summer the Sun would hit the bottom of a well in the Egyptian city of Syene (now Aswan), which meant that the Sun was directly overhead. Eratosthenes realized that if he measured the angle the Sun was casting in Alexandria (which was roughly due North) at the same time, it would be possible to work out the angle at the centre of the Earth between those two points.

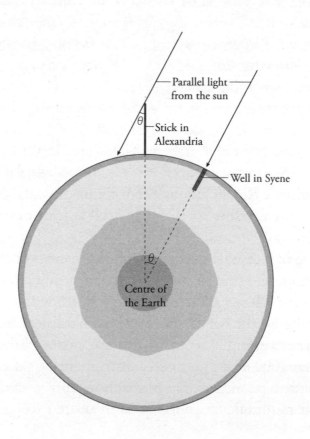

The distance that needed to be measured, in this case, is the one between Syene and Alexandria. We do not know if Eratosthenes sent people out specifically to measure the distance or if he used an off-the-shelf, accepted distance, but the consensus from later writers seems to be that he used a distance of 5,000 'stadiums'. Annoyingly, we do not know how long the ancient stadium unit was. People get as passionate arguing about its length as they do complaining that I use 'stadiums' as the plural instead of 'stadia'.

When Eratosthenes measured the shadow of a stick in Alexandria, the story goes, it was the equivalent of one fiftieth of a full circle, which we would measure as 7.2°. This angle revealed that the distance between Syene and Alexandria was roughly one fiftieth of the distance around the Earth. Multiplying the Syene–Alexandria distance of 5,000 stadiums by 50 gives an Earth circumference of 250,000 stadiums. However long that is.

I think worrying about the actual answer Eratosthenes got is missing the point. You can choose a value of a stadium to make the answer as accurate or inaccurate as you fancy. Plus the distance of 5,000 stadiums and the angle of exactly one fiftieth of a circle are suspiciously round numbers; I doubt they are what Eratosthenes actually measured. The point is that Eratosthenes's method was solid. With accurate input data, it would indeed spit out a pretty good measurement of the size of the Earth. The triangles were spot on.

You know I like to put my feet where my maths is, but I live in the UK which is too far up the planet for the Sun to reach the bottom of a well. So I decided to recreate a different iconic attempt to measure the size of the planet. Around the year 1000 BCE the prolific mathematician (also academic and writer) Abu Arrayhan Muhammad ibn Ahmad al-Biruni used a mountain in modern-day Pakistan to calculate the size of the Earth. He used the same method as Lucie and I did on our date: looking at the horizon. Specifically, al-Biruni wanted to measure what angle 'down' the horizon was.

On the beach we barely had to look down at all to see the horizon. Compared to looking directly straight ahead, our gaze was tiled by a mere 0.04234°, so it is understandable why, normally, we hardly notice it. From the top of a mountain, however, it would be a substantial angle (relatively speaking), and al-Biruni figured out that the angle you need to look down at the horizon is the same as the angle in the middle of the Earth between you and the horizon. He set about climbing a mountain, and my friend Hannah and I climbed up a building.

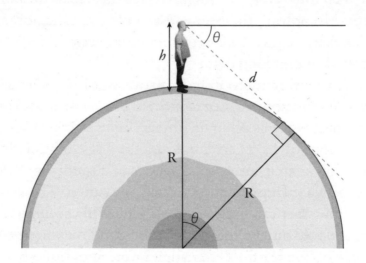

This is why we were measuring the height of the Shard back in Chapter One, the tallest building in the UK and our substitute mountain. My walk of pride in the 50-centimetre measuring shoes was the one distance measurement required by any calculation of how big the Earth is. Hannah and I both combined this with the angles we measured to get the height of the building. Interestingly, Hannah used tan values to get the height whereas I did it with sines. But thanks to the wonders of trigonometry we got the same answer.

Then we went up the building to get the angle down to the horizon. The Shard is a major tourist attraction so it is easy to book a trip right to the top and enjoy the view. The downside is that with public access comes private security. When Hannah and I showed up with a giant, home-made protractor and a large digital spirit level, the security staff decided they could be used to cause harm and confiscated our weapons of maths instruction.

Not to be defeated, we completed the trip to the top and estimated the angle down to the horizon using the inclinometers built into our smartphones which, as well as being hard to point accurately, only show results rounded to the nearest degree. We did our best and guessed the angle was about 1.5°. How wrong we were.

I say we were 'wrong', but we were right insofar as 1.5° is the angle we measured. Sure, once we used the cosine rule to get the ratio between the Earth's radius and the Earth's radius plus the height of the Shard, it indicated that the Earth is only 1,750 kilometres across. And, sure, modern-day nerds will have you believe that the Earth is around 12,750 kilometres across, but who are you going to believe? I went out there and measured it myself! Which I think makes me whatever the opposite of a flat-Earther is: someone who believes the Earth is a sphere but with even tighter curvature than NASA will admit.

As for al-Biruni's result, it runs into some common problems. The final answer he got from the mountain in Pakistan was a radius of 12,803,337 cubits. And of course we do not know exactly how long a cubit is. Classic. The variation in how big we think a cubit may have been covers what we now know is the exact radius of Earth, meaning the result could have been spot on (but probably wasn't). It's also worth noting that al-Biruni had at his disposal something very familiar to us by now: trig tables. This estimation of the size of the planet we live on was thanks to triangles and trig tables.

Any seriously accurate measurement, which came later in the second millennium, required a chain of triangles to get a really long baseline between two locations with a known angle between them. Willebrord Snellius of Snell's Law fame was an early pioneer in giant triangles and, in 1615, measured fourteen triangles chained across the Netherlands to get fairly close to the accurate answer. But Delambre and Méchain are the true champs, with their 115 triangles in the 1700s. They were measuring the Earth in order to establish the fancy new metre and so accuracy was their top priority.

Thankfully, technology had come a long way since the Romans were building roads. When planning Stane Street the Romans would have used something called a 'groma', which was effectively a bunch of plumb lines hanging from a cross shape. It allowed surveyors to continue a straight path by looking through the hanging strings and lining them up with points in the distance. It was only accurate to within human eyesight, and so there would be some random error in the measurements. The eyesight limitation was improved upon with the invention of the telescope, but removing other errors took some circular thinking.

The enemy of accurate measurements is random noise. If Delambre and Méchain wanted to measure the angle between two distance points they could point a telescope at each of them and measure the angle between the telescopes. But how accurately can you point a telescope at something? The human eye is still the hard limit on accuracy because we can only see so well. It might look bang-on through the telescope, but the alignment is bound to be slightly to the left or right of the true direction and that deviation is effectively random – that is, either direction is equally likely.

The antidote to random noise is averaging. Making a lot of repeat measurements and then averaging them will give you a more accurate result, even if the final number is not one you actually measured at all. We did a terrible version of this at the top of the Shard: it seemed our phones were saying 1° as often as they were saying 2°, so 1.5° it was.

Delambre and Méchain had a device with them to automate the process, as much as was possible at the time. Their two telescopes were mounted on a 'repeating circle', a disc that allows the telescopes to be rotated together. The repeating circle makes it easy to repeatedly measure the angle between two locations, and each result is added to a mechanical running total. The final total can be divided by the number of measurements to get an averaged and accurate result.

I have not tried to recreate the journey and measurements of Delambre and Méchain, but someone else has! In 2018 the Australian artist Sara Morawetz recreated the metre-defining walk from Dunkirk to Barcelona over a comparatively speedy 112 days, in a project called *étalon*. Sara does 'site-specific, durational performances', which is right up my street. I love a good bit of performance science art.

And the metre is crazy when you think about it. Humans live on a 1-metre scale – you can probably hold your hands about 1 metre apart – yet it is based on the size of the Earth, something which is incomprehensibly large to the human brain. However, if we walk far enough, taking steps which are roughly 1 metre long, we can gradually build up a sense of the connection between the human scale and the planetary scale. As we know, calculating the size of the biggest structure in the cosmos requires getting down and measuring a road.

Along the way Sara took measurements, and by the journey's end she had her estimation of the Earth circumference and was therefore able to calculate her own personal definition of 1 metre. To say I am envious is an understatement. About once per day Sara and her current walking companion would split up (she was joined by a succession of women artists as a gender-inverse of the scientific world at the time of Delambre and Méchain). One of them would stay put with a laser range finder and the other person would walk about 500 metres away, ready to be hit by said laser. By using GPS units they could also record the exact latitude and longitude of both ends of the measurement.

Obviously having GPS devices gives a huge advantage over the original measurement of the metre. Also, Sara et al. were sampling occasional 500-metre distances across this 10° wedge of the world, whereas Delambre and Méchain were going for an unbroken chain of triangles to measure the complete distance (which is also why it took them about twenty times as long, despite having more people). The mathematical effort needed, in a pre-computer age, to crunch the mesh of triangles is not to be sniffed at. Sara was able to compute each day's 'custom metre' as she went, and then work out an average at the end of the journey.

Her final custom metre was 100.038 centimetres, pretty close to the official metre of 100 centimetres! But the fact that Sara could record her metre in terms of, well, the metre, does raise the question: what units were Delambre and Méchain using to measure the original metre? The result of their final calculation was that the distance from the North Pole to the equator was 5,130,740 'toises', a pre-existing

French unit of length roughly equal to the arm span of a human. The metre was to be one ten-millionth of this distance, which would be 0.5130740 of a toise. This measurement was converted to lignes ('lines'), of which there are 864 in a toise (one arm-span was 6 feet, each foot was 12 inches of 12 lignes each: $6 \times 12 \times 12 = 864$). Thus, the metre was officially $0.5130740 \times 864 = 443.296$ lignes.

How close were Delambre and Méchain? Pretty close. We now know they slightly underestimated the polar circumference of the Earth: instead of 5,130,740 toises the equator–pole distance should be more like 5,131,766 toises. Technically the metre should be slightly longer, but once it was locked in place it would have been counterproductive to change it. Which is why the distance going all the way around the Earth, via both poles, is not 40,000 kilometres exactly but 40,008 kilometres. So close.

Stop. Haversine

When I asked Sara how the calculations for her custom metre were done, she replied that it was all done with haversine. Yes, what you thought was probably a joke trig ratio is actually super useful for calculating distances on the Earth!

I was once approached by a flight school based at Santa Monica Airport, just outside Los Angeles. One of the exercises they give their students is to try to take off and land in all 30 LA-area airports in the same day, and they were wondering what the shortest possible path was to achieve this. This is an example of the 'messenger problem' (often called the 'travelling salesperson problem') which involves finding the shortest path to visit some number of locations. Solving it is famously difficult, and beyond the scope of this book, but the first step was obvious: I needed to know the distance between each pair of airports.

I had a list of all 30 airports and the location of each one as latitude and longitude coordinates. This is subtly different to the x and y coordinates for, say, the locations of basketball shots on an NBA court which were recorded as distances. Those coordinates can be in metres or feet because basketball courts are flat. But the Earth is round, and that complicates things. For two locations far enough away on the Earth the shortest distance would technically require tunnelling directly between them. That is not helpful in almost any situation, particularly planning a flight route. I needed to find the distance around the surface of the spherical Earth. For that we use coordinates measured as angles.

Latitude and longitude have been around for over 2,000 years and still underpin modern GPS systems. The idea is simple: the Earth is a sphere,* and you can locate any point on a sphere by measuring two angles relative to its centre. Effectively, how far around (from −180° to 180°) combined with how far up (−90° to 90°). The angle of how far up, aka 'latitude', tells you how big a circle you move around on the Earth every day.

The Los Angeles International Airport (LAX) is at location latitude 33.9425°N and longitude 118.408°E which means it is 33.9425° up from the equator and 118.408° east of the prime meridian. Burbank airport is over at 34.2007°N, 118.3587°E, and calculating the distance between Burbank and LAX on the spherical Earth is not trivial. We need to

---

* The Earth is actually a spheroid, a shape a bit like a sphere. As an 'oblate spheroid' it can be considered a squashed sphere, which is a bit more bulgy around the middle. So the radius to the centre of the planet does actually vary slightly depending on where you are on the globe.

know the exact angle which would be formed if they were both connected to the very centre of the planet. If we know that angle, and the radius of the Earth, then we can easily work out how far apart the points are.

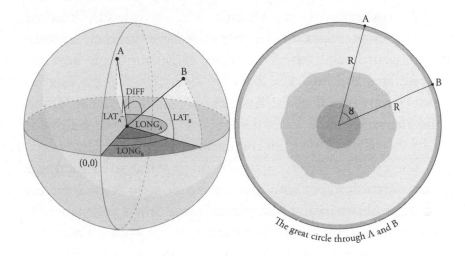

The great circle through A and B

Thankfully, haversine swoops in as a kind of Pythagoras proxy to work out that angle for us. The haversine of the difference angle is equal to a combination of the haversines of the latitudes and longitudes (with a few cosines thrown in for good measure).

$$\text{hav}(DIFF) = \text{hav}(LAT_A - LAT_B) + \cos(LAT_A) \times \cos(LAT_B) \times \text{hav}(LONG_A - LONG_B)$$

I was able to quickly write some computer code that used this haversine formula to spit out the spherical distance between every pair of airports in LA. I did also have to make some adjustments for directions of runways and restricted airspace, but the foundation of my calculations was haversine. Sure, you can build haversine out of sines but that is true of any trig ratio. The trick is to pick the best trigonometric function for the job and the reason we have haversine – one of the oldest of all the trig functions – is because it is so perfectly suited for calculating distances on a globe.

Sara was using it for the same reason, but in reverse. Instead of using haversine to get the angle between two places and then use the Earth's radius to calculate how far apart they are, Sara directly measured the distance between two places herself, used haversine to get the angle between them, and then worked back to deduce what the Earth's radius must be. Once she had her own value for the radius of the Earth, she could calculate the circumference and divide it by 40 million to get her own value of the metre.

You don't have to recreate the original trans-France metre measurement to get your own metre. Somehow measure the distance between any two places, note down their latitudes and longitudes, and use haversines to get your own metre. Note that a personal metre is pronounced *me*tre.

An Important Point

Where would we be without triangles? Lost, probably. As well as the definition of the metre, triangulation has given us centuries of accurate maps, only recently superseded by aerial photography. The signs of these triangles still cover the landscape, if you look closely.

The ancient landscape of Great Britain is littered with ancient monuments and standing stones, but some of my favourite rock creations are altar-like pillars erected within the last century. They look like they are from a bygone era, but these 'Triangulation Stations' were erected between 1936 and 1962 as part of something called the 'great retriangulation of Great Britain'. Now obsolete, they still dot the landscape.

Matt Parker @standupmaths · May 9
I do enjoy a good Triangulation Station.

Of course, people try to locate and visit as many as they can, an activity called 'trig pointing'. The website trigpointing.uk has 6,871 trig-point locations logged. Certainly, whenever I encounter one I cannot resist documenting my visit with a few quick photos: at a minimum, a shot of the trig point's ID number and one photo of me grinning like an idiot in front of it. From each trig point, it was originally possible to see at least two other trig points, usually on distant hills. Which means, as hobbies go, they are almost always in scenic locations.

For the lay rambler it is not immediately obvious what a trig point is. On the top there is a cryptic, three-pointed asterisk shape. While the channels making up the pattern look like something from the climactic scene of the 1998 Wesley Snipes film *Blade* (that's a niche reference), they are actually designed to hold a theodolite, a surveying device contemporary with the repeating circle. In my experience, most people will take a closer look at a trig point when they stumble across one, get confused by its hieroglyph-like markings, and then walk on, none the wiser they have just encountered the basis of our modern maps.

-0.848048

The precursors to these trig points were from an earlier triangulation of England which started about the same time as Delambre and Méchain set off, but took about half a century longer (218 points were measured between 1783 and 1853). This was the 'principal triangulation' of Great Britain and resulted in the first publication of Ordnance Survey maps. Still published to this day, they are affectionately known as OS maps and, despite them now having a perfectly good app, I must confess I also have a drawer full of paper copies for old-school walking holidays. Details of every footpath, bridleway, building and everything else a rambler needs to know about has been reduced to 4 centimetres to the kilometre, with contour lines so you know how steep your stroll will be.

Originally planned for military use, when OS maps were published in the early 1800s they were an instant hit with the general public, despite costing over a week's earnings for most people. They were the only way to get a birds-eye view of the land (excluding our friends the hot-air balloons, which were absolutely no less terrifying in the past).

These early OS maps also, for the first time, allowed people to know for certain where everything was. It's reported that before the principal triangulation everyone assumed that Cape Cornwall (in, well, Cornwall) was the most westerly point in England. I've been there, and it definitely looks like what you want the end of a country to look like: a piece of land with dramatic cliffs jutting out into the ocean. But a few miles south is the actual most westerly point, Land's End. Much less dramatic, it is more of a bump in the coastline. Standing on Cape Cornwall, I got my compass out, pointed it down the coast at Land's End and could not convince myself it stuck out more. Whereas my trusty OS map – based on unambiguous triangles – clearly revealed where the land actually ended.

-0.857167

By the 1930s, it was clear that the principal triangulation had not been clear enough. A new wave of triangles swept across the country between 1935 and 1962 in the great retriangulation of Great Britain. This was interrupted by the Second World War, which required the production of 342 million maps (120 million just for the Normandy Landings), but by the 1960s detailed maps covered everywhere in England, Scotland and Wales.

Even before the great retriangulation was over, the end of trig points was already on the horizon. In the later years, on-the-ground surveying was being supplemented by aerial photography. In the modern era, between planes and satellites, trig points are no longer needed and the Ordnance Survey has stopped maintaining them. I dug through their archived data, and the last maintenance conducted was on the 10 October 2002 when two trig points, Cnoc Moy and Meall nan Con in the Scottish mountains, were serviced by some intrepid trig-point technicians.

The direct descendants of the trig points are a network of about 115 GPS receivers, which are fixed in place around Great Britain so that most of its area is within 75 kilometres of at least one station. They constantly record GPS data and once an hour it is all collected together just to make sure England, Scotland and Wales are all where we think they are. Our modern navigation triangles are digital and automated, while the triangulation stations have become yet more historic bits of stone littering the countryside.

The UK is not alone in having the marks of surveys past etched into the landscape. Even in the middle of New York City there are signs of historic surveying. When the current grid system of streets was being planned in the early 1800s, it required mapping out exactly where all the street intersections would be; these places were often marked by driving a

-0.866025

metal bolt into the ground. Not everyone was keen on this idea, as it involved displacing whole communities and those affected would swiftly remove the markers as soon as they were put in place. But the plan did go ahead and Manhattan was covered in precisely planted metal bolts.

The precision of the Manhattan grid is testament to the triangulation accuracy of John Randel, who knocked the bolts into the ground. Unlike trig stations, which were built to last, the bolts were removed once the roads were in place. Even so, at least one Randel bolt remains. I know because I have visited it! In 2004, professor of geography Reuben Rose-Redwood and surveyor J. R. Lemuel Morrison set out to see if any were still extant. Their theory was that, since not all of Randel's roads ended up in the final plan, the locations of those 'phantom roads' in modern-day parks might still have some bolts.

Sure enough, in the middle of Central Park there is one such bolt. Its location is a loosely guarded secret, ostensibly to protect it from too many people interrupting its two-century slumber. But in reality, anyone with enough time, effort and internet-sleuthing skills can find it. I'm proof of that. To keep the challenge alive for people who also want to try and find it, I'm going to maintain the tradition of not revealing its location. And, if you want to play this game on hard mode, there are rumours of as many as two other original bolts somewhere in Manhattan, but I have not found any of them myself.

Where are We Now?

We are in the future now and the future is all about space. Forget physical maps based on actual trig points: now we have online maps, made using satellites in orbit. If you want a system for finding your position globally, you need GPS.

-0.874620

GPS has been mentioned a few times in this book already, but 'GPS', for 'Global Positioning System', is actually a brand name. There is more than one system of spacecraft beaming precise radio signals to form a global-navigation satellite system, giving us the generic name GNSS. The OS actually describe their 115 receivers as 'geodetic GNSS receivers', but this is a Kleenex/Hoover situation where the GPS brand-name has become a generic term in the eyes of the public. The OS have also called their sky-facing system OS Net which is a bit too Terminator-y for my liking.

The actual GPS belongs to the US military. Out of the goodness of their famously big hearts, they allow a free civilian version of GPS so anyone can measure their position, anywhere on the planet. This was originally restricted to a lower resolution, but since the year 2000 the US military claims the free version of GPS is as good as the military one if you have the best equipment. In theory, this means anyone can know exactly where they are within centimetres, maybe even millimetres if you stay still and collect data for long enough (like the OS Net receivers do). But is it true the US military does not keep a better version of GPS for themselves? They have for satellite imagery, and they could very well have for coordinates too. We may not know until some future president shares a classified photo on social media.

Whatever the GNSS, it uses the same old triangles we already know and love. But, instead of using fixed trig points cast in stone, the points have been cast into orbit. Each satellite beams back very accurately time-stamped radio signals, which can be used by a GNSS receiver to triangulate the relative distance to any number of satellites whose positions are known in great detail. Like an Earth-sized version of the 'two trains leave their stations at the same time' problem, but with photons instead of trains and we actually care about the answer.

The satellite-navigation revolution has been hugely impactful, providing easy navigation and location information to anyone who can afford a GPS device. At the push of a button anyone can know their exact latitude and longitude on the Earth's surface incredibly accurately. This has done much for the betterment of humankind but has also provided us nerds with yet another hobby: GPS coordinate hunting. The Degree Confluence project was started almost as soon as hand-held GPS devices for hiking became available. Founder Alex Jarrett says it began because, in 1995, 'my friend managed to convince me to buy a GPS and I had to come up with something to do with it'.

Alex decided he liked whole numbers and set about trying to find and document integer crossing points of latitude and longitude. The first place he visited was 43.00000°N 72.00000°W, an arbitrary location in New Hampshire, USA. After he set up the website confluence.org, volunteers joined in with an 'organized sampling of the world'. The goal was to visit every whole-number crossing point of latitude and longitude on land which, once a bunch of points near the poles were excluded, was a total of 9,704 spots to be visited. Each visit requires the GPS adventurer to take one photo of the spot on the ground and photos facing north, south, east and west.

I loved this idea of an arbitrary but systematic sampling of the planet so, in 2004, I decided to join in. After two days of hiking through the Australian desert with some friends, I became the first human to visit and document the exact crossing point of latitude 26°S with longitude 115°E. The precise location looked just like the rest of the red-dirt bushland we'd spent several hours hiking through, but we knew it was special. We took the required photos and then began the long hike back.

A very precise but otherwise unremarkable location.

That long hike through the Australian wilderness, 17 kilometres from our car which was itself a 110-kilometre drive from the nearest proper road (and then hundreds of kilometres to an actual town), was a startling reminder of why GPS can be so important. On several occasions we were so disorientated in the homogenous, flat bushland that my instinct was telling me to walk in a completely opposite direction to what the hiking GPS was saying. As strange as it was to feel like I was walking away from the car, my logic held for several hours and sure enough we emerged exactly where we had left our supplies. I can say with absolute confidence that, if it were not for triangles in space, I would not have survived that adventure.

If you would also like to be the first to visit a whole-number latitude and longitude crossing point, at the time of writing there are still 384 unvisited. Given they are even less accessible than the one I had to hike to through the desert 20 years ago, it is probably best to revisit one of the already-documented points. A second goal of the project is to document the changes over time at these locations, so a revisit is never a waste. You're never more than 80 kilometres from one of these confluences of latitude and longitude lines.

-0.898794

The GNSS revolution has also exposed the mild inaccuracies of the old ways of measuring things. One of the touristy things to do in London is to go to the Greenwich observatory and stand on the 0° line of longitude. The Earth's spin gives us a clear equator right in the middle of the planet, and the axis of rotation gives us precise north and south poles, so angles of latitude have an unambiguous reference frame. Longitude, however, which measures how far 'around' the planet a location is, has no such obvious starting point. For a while some countries just used their own starting points, but that was exactly as confusing as you would expect.

So in 1884 representatives from twenty-five countries got together in Washington, DC and voted on it. Greenwich Observatory in London won with twenty-two of the votes (the dissenting countries were Brazil, France and San Domingo). This designated the starting point, or 'prime meridian', as the line which started at the Greenwich Observatory and then went all the way around the planet, through both of the poles. The line is actually a massive circle going exactly around the very centre of Earth. Or it would have been, if they hadn't missed.

To draw a circle which goes around the centre of the planet you need to know exactly which way is down. This was done very carefully using gravity to indicate the direction 'down', but the intrepid surveyors did not know that the density and deformities below the surface of the Earth could cause local gravity to misalign with the geometric centre of the Earth. Thus the original 'zeroth' circle around the Earth wasn't actually centred on the middle of the Earth. Nobody noticed at the time, though, and a big shiny metal line was put in the ground at Greenwich for tourists to jump back and forth over.

Satellites have no such problem locating the middle of the planet and so have a more accurate 'down'. With the global adoption of GNSS there was now a mismatch between the true longitudes and those based on the metal line in Greenwich. In 1984, exactly a century after the first gathering, countries from around the planet got together and decided on a new prime meridian, ever so slightly to the side of the old one.

Until GPS devices became affordable this had no impact on the tourists in London, but now that anyone with a smartphone can see their exact latitude and longitude, waves of tourists get very confused why they don't see a reading of 0° when standing on the famous zero line. If they do follow the numbers on their phone they will reach 0° about 100 metres off to the side. Which is a bit less crowded and makes taking your photo easier, but now, instead of standing on an iconic line, you're standing on some grass in a park.

The messiness of the physical Earth, and the precision of modern latitude and longitude, is also why we need to have three different norths. The classic north you are probably thinking of is the direction up the lines of longitude directly towards the North Pole, as defined by the rotational axis of the planet. This is called True North and in a perfect world it would be the only north we need. The second north is caused by our insistence on having maps. I used haversines to calculate the distances between airports because I know the greater LA area is actually a section of a sphere, but if you get a physical map of LA it will be on a flat piece of paper. It is the tension between the curved Earth and flat maps which gives us Grid North.

Every line of longitude goes through the North Pole, meaning they all cross at that one point. While lines of latitude are nicely spaced, lines of longitude get closer together as they go from the equator to a pole. This is why the Degree Confluence Project removed a bunch of crossing points near the poles as destinations – because there are so many of them, all packed quite close together. It is also at odds with an organization like the Ordnance Survey who want their maps to have a neat, square grid where the lines do not bunch together and the scale on the map doesn't change from place to place. So map-makers invent their own north.

The OS decided that their Grid North would match True North along the 2°W line (the whole-number line of longitude closest to the middle of Great Britain). Because Grid North lines are parallel and True North lines all curve together, the further you are from the 2°W line the more True North as shown on your GPS device will deviate from Grid North shown on your map. However, if you get a compass out, it will probably give you a whole new direction for north. Magnetic North is the third type of north because the north and south magnetic poles do not align with the rotational axis poles. And, to make matters worse, they move around. Moreover, the lines of Magnetic North are not even straight.

The Earth has a magnetic field, but it is not as neat as the sort of magnetic field often depicted around a bar magnet. Like gravity, the Earth's magnetic field depends on the density and composition of what is inside the planet, and this means that the field lines bend and move all over the place. All a compass is doing is showing you which way the local magnetic field is pointing, which is often 'good enough' for navigating but compared to GPS, or an accurate map, the difference can become noticeable.

There are some places where, just by chance, the local, distorted magnetic fields sometimes happen to line up with True North. Due to the turmoil deep within our planet, these points move around as the magnetic field slowly flexes. There has been one such line across the UK: if you stand anywhere on that line, your compass will point directly at True North, and that line has been gently drifting towards the west. In November 2022 the magnetic field shifted far enough that it overlapped with the south-most point in Great Britain on the 2°W line.

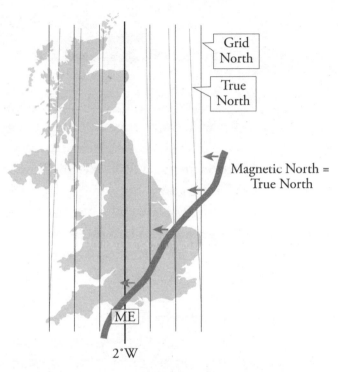

As the magnetic zero line drifts to the left, the triple-alignment will move up the 2° line.

Of course I set off for a hike with an OS map, a magnetic compass and a GPS device to find the one spot where, for the first time in known history, all three norths aligned in England. Standing on a windswept cliff above a winter ocean, I was the only person in Great Britain who had no ambiguity

That's my north face.

about which way was north. Since then, the single alignment of three norths has drifted slowly, well, north. When this book is published I predict the Three North Point will be roughly on the M62 motorway between Manchester and Huddersfield. Then, sometime around July 2026, it will drift off the north coast of Scotland, back over the ocean, unlikely to return this millennium. As the person who welcomed the alignment to our shores, I'll be there to wave it goodbye.

Relativity Right

There is one last way in which triangles allow us to know our location on Planet Earth using modern GPS, and it involves understanding the very shape of reality around us. Modern physics comprehends the universe as existing in a 4D structure known as 'spacetime'. Much as the distortions of a 2D surface determine what shapes you can draw on it, the distortion of 4D spacetime changes the shape of objects around us, and even the way time passes for them.

I mentioned earlier in this chapter that GNSS works because

the satellites beam out 'very accurately time-stamped radio signals' but, because of the effects of Einstein's relativity, the satellites are experiencing time differently to you and I down here on the surface of the Earth. From our perspective, the satellites are set to the 'wrong' time and that will throw all of the distance calculations off. Things like GPS rely on knowing how to calculate the exact time-distortion between you and the satellites, and this requires some triangles which are much less complicated than most people would expect.

Einstein's work has the aura of being un-understandable, but it really only requires the Pythagorean Theorem and a bit of thinking. Einstein imagined someone on a train measuring time by bouncing a photon between two mirrors, one on the ceiling and one on the floor. If you were along for the ride with the photon you would see it move a distance, 'd', from the roof to floor. But for people outside, watching the photon move as the vehicle races by, it will appear to travel a longer path as the train flies by carrying the light-clock with it.

For anything other than a photon, people outside the train will also see it moving faster as the train carries it along. But the unexpected thing about photons is that they always appear to move at the same speed no matter how fast you are moving relative to them. (This is different from light slowing down when it goes through glass, water or the atmosphere. This is technically only in a vacuum.) If you were floating in space, and someone threw a torch towards you at half the speed of light, your last thought would be, 'Huh, the photons from that torch should be going at 1.5 times the speed of light, but they still look to me like they are only going the speed of light.' If you could really think fast.

Einstein realized this would affect the passing of time if the universe actually behaved that way. I've drawn our photon-on-the-train from the points of view of someone on

-0.939693

board and someone watching it whizz by. On the train, the photon takes time 't_1' to go from the roof to the floor, and for anyone watching outside it takes 't_2'. For anything not a photon, $t_1 = t_2$ but in this case the photon always goes the speed of light. I've used the Pythagorean Theorem to calculate the sides of our photon triangle and get the ratio between t_1 and t_2 which is called the amount of 'time dilation'.

Most of the time, this kind of hypothetical working-out of a thought experiment turns out to be completely theoretical, or at least does not apply to the universe we live in. But Einstein was right on the money. This *is* how our universe behaves! If something is moving it will experience time passing at a slower rate than a comparable object not moving. Scientists have even launched very accurate clocks into space (where it is easy to move fast) and this time dilation has been measured.

The reality is, time on the satellites progresses at a very slightly different rate than it does for us on the ground. Like a really boring version of the film *Interstellar*. But, also like in the film *Interstellar*, there is more than one type of time dilation. Einstein was not a one-hit wonder and, after dropping Special Relativity as the biggest physics hit of 1905, he came back with 1915's smash sensation, General Relativity. These equations are a bit more complicated, but the cheat-sheet version is that gravity also changes the rate at which time passes. Bizarrely, if you happen to be near something with a lot of mass, you will start to experience time at a very slightly different rate.

The final twist is that gravity-based time dilation cancels out the velocity-based time dilation experienced on a GPS satellite. The satellites in question are moving fast, about 14,000km/hour, which means that one complete day on the satellite is 7.2 microseconds shorter than what we experience

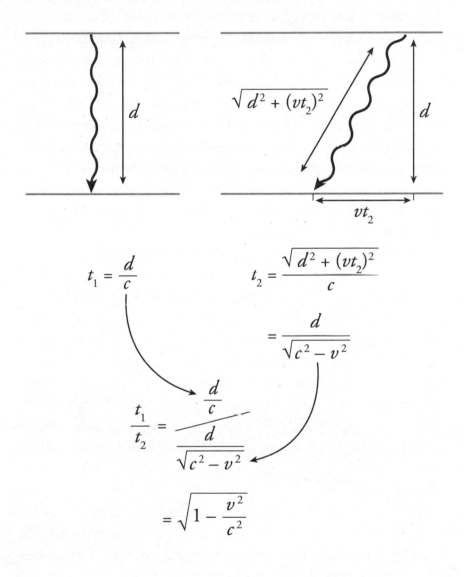

here on Earth. However, we are closer to the mass of the Earth which, in turn, slows down our time dilation to the tune of 45.6 microseconds per day. This offsets the 7.2 microseconds caused by our speed difference and leaves a residual 38.4 microseconds per day difference, with time passing 'too fast' on the satellites.

Because we understand the photon triangle (and the other General Relativity calculations), we are able to compensate for time dilation, and this means GNSS is possible. Without this abstract physics and trigonometry, systems like OS Net wouldn't be able to achieve millimetre-accuracy when measuring a country. And I wouldn't be able to risk life and limb hiking to interesting latitude and longitude locations.

Nine

BUT IS IT ART?

Enough with the applications! It's nice to have GPS to make sure we don't get lost, but what if we want to use triangles and trigonometry for something arty? We need to swing back in the geometry-for-fun direction and look at some art. I'm not talking about 'maths is art' in the abstract, even though I agree it is and it's delightful in its own right. And this chapter isn't going to be about using trigonometric functions to make maths-art like fractals (don't get me wrong: I could go on about fractals for ever). Instead, let's take a good hard look at how triangles have enabled art as we know it today. Specifically, how can we use trigonometry to create 2D depictions of the 3D world around us?

We will start with one of the more mechanical forms of art: photography. I think of photography as something like a 'reverse torch'. Imagine having a point-source of light, like a lightbulb, which is blasting photons, moving in straight lines, onto everything in a room. A photograph is the reverse: it's all the light coming back from everything in a scene and being captured.

The trick to photography is making sure only the correct photons are selected. If you just held a digital camera's sensor (or old-school photographic film) up in the air, it would collect photons but the resulting image would be a washed-out blur with no discernible features. This is because the sensor would be hit by photons coming from all over the place. To get a focused image, you need a way to select only the photons coming from the correct direction. The crudest way to do this is with a tiny pinhole.

Let's say we want to take a photo of a red balloon. But that balloon is blasting photons in every direction! A pinhole camera filters out unwanted photons by having a single, tiny hole; only the light which hits that particular point makes it through. Each part of the balloon aligns with a different part of the sensor and so we get a sharp image. The geometry of object–hole–sensor is also why images from pinhole cameras (indeed, all cameras) are upside down.

Some animals have pinhole-style eyes, such as snails and octopuses, but we humans are one of the many animals who evolved a lens to focus the light. Filtering out so many photons is a bit wasteful and, because so little light gets through the hole, it can take a very long time to expose a good image. Better to take multiple photons, which all originated from the same point on the balloon, and redirect them all onto the same point on the sensor. This is what lenses do, enabling us to take photos without having to sit perfectly still for hours.

As well as evolving lenses in our faces, animals have also had to evolve brains that can process visual information. We may perceive the world in 3D, but our vision has gone through a 2D projection bottleneck and our brains need to compensate for that missing information. In 2D not everything looks the way it looks.

If you stare directly at a picture of a circle, it will look like a circle. Tilt that circle over a bit and your brain will probably still tell you it's a circle, but the actual image being projected

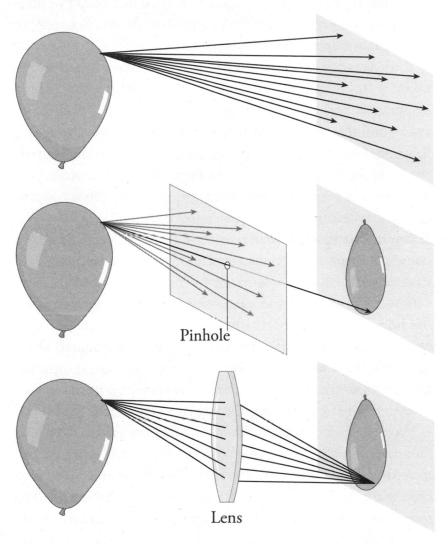

Pinhole

Lens

Many rays: bad. One ray: good. Lots of rays on the same point: better and brighter.

onto your retina is no longer a circle. It's now an ellipse. Trump's photo of the Iranian launchpad was in this situation. The pad itself was a circle but the satellite taking the image was off to the side, so the circular launchpad was projected as an ellipse on its sensor. It was the shape of this ellipse which allowed people to calculate the angle to where the satellite must have been looking from.

The distance an object is from the camera also changes how it looks thanks to an effect known as 'perspective', which sometimes makes it possible to calculate exactly how far away something is in a photograph. In a very simple sense, as things get further away they look smaller. And the maths is also simple: if something is twice as far away, it will look twice as small. You can test this yourself: a 2p coin is roughly one quarter bigger than a 1p coin (by width, 27.6 per cent) so if you hold a 1p coin 40 centimetres from your face and a 2p coin 50 centimetres – an extra 25 per cent – away, they will look exactly the same size from your point of view.

This is the basis of the calculations that were done to work out how high the hot-air balloon was above the pigs. In that case there were a bunch of added complications because of the optics of the lens on a camera. I do not want to get derailed talking about focal distances, but there is one interesting thing to pull out: I later said that the equation used an angle to calculate the height of the balloon. That is because an efficient measurement for capturing how big things look is 'angular size'.

A 1p coin at 40 centimetres and a 2p coin at 50 centimetres have the same angular size. If you want to look from the top to the bottom of each coin, your eyes need to rotate down the same angle (in this case, about 3°). In the same way, as a person or object moves towards you their angular size will gradually increase until they are technically 'all up in your grill' (aka an angular size of 180°, not that your eyes can open that wide).

Angular size is a relative measure of how things look, separate to their physical size. If we have some kind of reference measurement we can use careful calculations to go from one to the other, like taking the 34.5° size of the Space Orb and working out how big it actually is, or calculating the distance of the hot-air balloon above the pigs. Our brains, however, are usually in a rush and forgo length calculations in favour of just having a guess.

Even though a distant object is projecting a tiny image onto our retina, we'll still perceive it as being its actual size. The size of an image on our retina will give our brain a measure of its angular size, and the brain then autocorrects that to a rough actual size. Our visual system effectively compensates for how far away an object is and does not bother to keep us in the loop. You may feel like you are viewing object reality, but your subconscious brain is doing a bunch of hidden guesswork.

These Matts are exactly the same size!

-0.978148

Here's a challenge: try to turn off your brain's autocorrect. In the photo on the previous page, I've edited in a second, giant copy of me, chasing small me. In the image the two Matts appear to be different distances away, and therefore your brain is screaming at you that the 'distant' one is substantially bigger. But on the actual page, they are exactly the same size. They have the same angular size, and they are both making exactly the same size projection on your retina. It is everything your brain is doing after that which makes you think they are different sizes.

New challenge: I've photographed myself twice in the same location. You're staring right at Close Matt and Distant Matt. I have not altered the photos at all and did not grow any taller between taking them.

Trying to find myself.

-0.981627

In both photos I am one 'metric Matt' tall. Because Closer Matt is closer, he is of course bigger in the physical image before you. Without measuring it, how much bigger do you think the 'closer' version of me is? At least decide if Closer Matt is bigger or smaller than two times Distant Matt. The answer is below.

I talk a lot about how big things are when they are 'projected onto your retina' because that is both how we perceive the world and a really useful mathematical way to understand it. You're watching a little movie in your head, with your retina being the screen, your brain the audience and the whole world a stage. But your brain (and whole visual system) does a lot of processing to get around the 2D bottleneck. The upside is that we can make works of art which fool your

Psych! Closer Matt is over four times the man Distant Matt will ever be.

−0.984808

brain into thinking it is looking at a scene, when in fact all it can see is paint on a canvas.

Paint by Numbers

This is not a book about art and I don't want to get too drawn in to discussion about artistic licence and whatnot. But I do want to say that I appreciate that art is not all about accurately recreating exactly what humans see. It's art!

Art has no obligation to depict reality, and the cultural requirements of art across the Earth and throughout history have varied wildly. But I would like to take a moment to look at European art's attempts to show buildings in a way which I can best describe as 'not unsettling'. When a culture lives in a built environment of straight lines and right angles, artists depicting that setting need to draw off-angle lines in a way that makes them 'look right'.

This mosaic dates from between roughly 100 BCE and exactly 79 BCE. We know it was definitely made before 79 BCE because it was installed in a Pompeii villa which was volcanoed in that precise year. It shows Plato's school in Athens (which looks a lot different from when I visited it), though we don't know for sure which of the seven people is Plato. The most likely candidate is the individual third from the left, pointing at the ball with a stick. However, it is the individual third from the right, directly aligned with the solitary column, that I most identify with, since he's making the same facial expression I did when I first saw this mosaic: mathematical confusion.

For a start, the seven people are clearly supposed to be arranged in 3D space but it's . . . not very convincing. They all just kind of float there in 2D. There is some use of light and shadows on the ground, to indicate where the two box-things are relative to each other, but it's all very implausible. I know that this is a mosaic, and it feels a bit cheap to be dunking on an inherently low-resolution artform, but it definitely demonstrates the point I want to make. Plus, I find it deeply pleasing to use an image depicting one of the greatest and most influential minds in geometry to showcase bad geometry. Third-from-the-right knows what I'm talking about.

Hmmmmmmmm = Hm^7.

-0.990268

Part of what I find so disorientating is that things do not get smaller as they get further away in the image. All the people are roughly the same size, even though they are intended to be sitting different distances from the viewer. Which can be fine, of course. An artist may not want to use the size of figures to represent physical location. Throughout history the size of figures in an artwork has been used to indicate all sorts of things, such as their relative importance. But here it feels like they want to be sitting around in 3D.

Then there is the archway in the top left. The left-hand column is clearly a greater distance from us than the one on the right, but they are both the same width in the mosaic. And the artist seems to know that the top will appear to be on an angle from our looking-up point of view, but it doesn't taper away as it moves back.

I say 'taper away' like it's obvious what should happen. It is absolutely not. Even if you have an understanding that objects in the distance appear smaller, and that otherwise-straight objects appear to be on an angle when viewed from different viewpoints, there is no easy way to paint all of that at once in a systematic way without mathematics. This leads, in the Pompeiian mosaic as in many other artworks, to a patchwork of different alignments. 'Empirical perspective' is the catch-all name given to when an artist slaps together whatever type of perspective works in each different section of the painting.

It took a while before someone realized that empirical perspective was not the best option if you wanted a painting to look natural and the viewer not be distracted by people floating in 2D. What I find amazing is that it's not like artists everywhere had their own epiphanies for how to draw with true perspective. It was discovered by a single human: Leon Battista Alberti. From him, the information flowed out like a virus of pure knowledge until pretty much all of Europe was painting in perspective.

Alberti worked for the Catholic church in the 1400s, and one of his other important works was a book about cutting-edge cryptography. If anyone has wanted to keep this triangle knowledge a secret, he would be the guy. I want to argue that what Alberti discovered was so obvious that there was no use trying to hide it, but that begs the question of why no one else independently uncovered it. Thankfully his perspective work was put in a book called *On Painting* and distributed widely with nothing to hide. He saw what no one else could see: the vanishing point.

Yes, Alberti's discovery can be summarized in a single point. The vanishing point. He realized that if things get visibly smaller as they get further away, then everything must converge together into a singular singularity. In a painting, he discovered, everything must be drawn in like it was being drawn in.

Which is not to say that people were not getting close. In 1305, the Italian artist Giotto got real close with a ceiling. The beams certainly look like they are closing in on a vanishing point. But an analysis by the vision expert Christopher Tyler showed that they do not converge in a systematic way. They are all close-but-not-quite, implying that, while Giotto had the right idea, he was doing it by eye. Moreover, the steps and everything else in the painting do not show any indication of perspective convergence. When objects like the steps are drawn with straight lines, but the lines show no effects of perspective (they stay perfectly parallel), it is called 'isometric'. So Giotto painted an isometric room and then added a really great, perspective-convergent ceiling.

The Alberti style of painting involved putting a single dot on the horizon and then using a ruler (or similar) to draw

Jesus Before the Caïf by Giotto (1305), from Perspective as a Geometric Tool that Launched the Renaissance by Christopher Tyler (2000).

straight perspective lines radiating from that point. This technique took all of the guesswork out of drawing and meant that every section of the image had the same perspective. It was a systematic method to make paintings which looked, to humans, the same way reality looks.

So when Raphael got ready to paint *The School of Athens* in 1509, not only could he bring Plato inside but he could make sure all of the geometry made sense. At last, Plato could point upward with a sense of, 'Check out these arches, now that's how you depict perspective!' And it doesn't get more Renaissance than this painting. While Raphael was applying the paint to plaster in the Vatican, a few rooms down Michelangelo was slapping paint all over the Sistine Chapel, and Raphael's Plato is a suspicious lookalike of his hero Leonardo da Vinci. Actually, Raphael hid a lot of double meanings in the painting, which is how we know his favourite band was Guns N' Roses: he hid the cover of their 1991 releases *Use Your Illusion* in the mid-right section of the painting.

That is definitely not what I remember the park looking like.
But it's undeniably lifelike.

And Raphael was using an illusion. Alberti's linear per-spective is actually a form of trickery: it recreates only a single viewpoint, and the illusion is broken if you view the painting from any other location. It happens that putting the vanishing point in the middle of a horizon, about two-thirds of the way up the painting, places it about where someone looking at the painting is likely to have their eyes. There is a decent tolerance, so standing anywhere roughly opposite the vanishing point makes the painting look realistic.

But some artists would roam the vanishing point all over the place, sometimes even out of the bounds of the painting. As such, if someone stood next to the painting, staring at the wall adjacent to the artwork, the depicted image would look correct in their peripheral vision. But looked at dead-on it would be weird.

The problem is depicting boxy objects like buildings. A box actually has six vanishing points, one in every direction

Just a box minding its own isometric business, looking at the horizon.

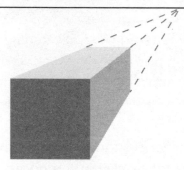

Same box, but with a single-point perspective centred on the horizon.

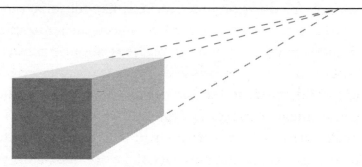

Oh no, the vanishing point has moved too far to the side.

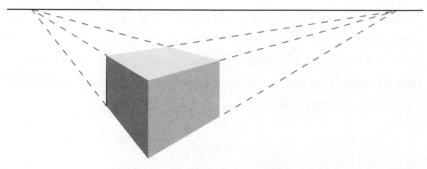

A second vanishing point appears to save the day!

its faces face. The vanishing points directly above and below do not matter for most paintings as the view is looking straight ahead, and not up into the sky. If one face of a boxy building is dead-on to the point of view of the painting, then a single vanishing point in the distance is fine. The opposite one is behind the viewer and the two side ones are, well, off to the sides. But if a box is viewed from a bit of an angle, now there are technically two vanishing points visible to the viewer, and everything will look a little off if the artist sticks religiously to just the one vanishing point.

Despite the single-vanishing-point style catching on from the 1400s onward, there is not a single example of a two-vanishing-point painting before the 1600s. Maybe it's not as obvious an idea as it seems to us, with our advantage of hindsight.

This illusion problem was neatly avoided in many cultures by not worrying about it. Some contemporary Chinese paintings were only a metre or so high and many metres wide. For a painting in such extreme widescreen there is no one canonical viewpoint. Someone needs to be able to walk the length of the painting, and having a single perspective would look ridiculous. So, as well as artistic license, there are solid, practical reasons why an artist would want to avoid linear perspective and use regular, reliable isometric perspective instead. Isometric still lines everything up neatly, but keeps the lines parallel instead of converging on a vanishing point.

Now that photography is taking some of the heavy lifting when it comes to straight-up documentation, Western artists are once again throwing off the shackles of linear perspective. These days, our mathematical understanding of perspective and projections means we can ignore reality and make some very trippy art.

Beauty is in the Eye-Line of the Beholder

The next time you are driving down a highway, have a guess at how long the dashed lines on the road are. Or, given you're now a member of the elite club of informed people who have read this book, ask someone else in the car. Studies conducted in the USA show that most people will say the dashed lines are about 2 feet long. Everyone agrees the lines are shorter than an adult human. But in fact US federal regulations mandate a dashed-line length of 10 feet (and it used to be 15 feet). Those lines are looooong.

These lines need to be clocked by drivers moving at high speeds, looking at the road on a pretty oblique angle: they have to be long to appear to be short. Similarly, if you have a look at writing on a road which was designed to be read by motorists: it's incredibly stretched out. Likewise, a bike symbol to indicate a cycle lane will appear to have circular wheels, but looked at up close they are stretched-out ellipses.

Circle from the front, ellipse from above.

-0.999848

This is an example of anamorphic art, which is artwork (or, in this case, a functional graphic) designed to be viewed from one specific viewing location. The image is painted to look like a bike from the driver's seat of any approaching metal death-boxes. Anamorphic art takes advantage of the ambiguity in human vision. Your brain takes the light detected by your retina, and has to decide what are the most plausible surroundings to have caused that pattern. Instead of thinking in terms of my retinas, I like to imagine a floating frame in front of me which is a snapshot of what I'm seeing. Any combination of light which makes a sensible image on that plane will be interpreted by my brain as such.

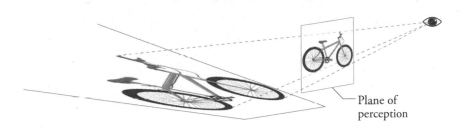

Plane of perception

A stretched-out bike on the road looks like a normal bike floating in the air.

This method can be used to paint 3D-looking patterns on a flat road. Cycling through south London recently, I noticed that Lambeth Council was saving some money by painting illusion speed bumps instead of actually building them. Or maybe it was an enterprising local who had plenty of paint but no construction materials. Whichever the case, the question is: do they actually fool anyone? Of course, once locals know the road is flat they will have no actual need to slow down, but the illusion might be enough to remind them to. This idea has been tried in other countries before, with

Iceland painting in anamorphic objects designed to make drivers slow down out of sheer confusion.

Fake speed hump in south London, which causes the locals to pretend to slow down.

If you speed in Iceland you face serious prism time.

-0.999848

Behind the scenes, the geometry is no more complicated than anything we have seen so far. Light moves in straight lines and using trigonometry it's possible to take a line between where the viewer's eye will be and the intended image on a plane of perception, and calculate where an extended version of that line will hit the artwork surface. The calculations are straightforward enough that they can be automated and performed 60 times a second to insert anamorphic images into a live video-stream.

Anyone who watches American football will be familiar with the magical first-down line that appears in TV broadcasts of the game. During a game, the offensive team needs to advance at least ten yards over the course of four plays, but this target – the 'first-down line' – is relative to where play started and changes constantly during the game. Players on the field can keep track of where it is (or look for markers on the side of the field) but that's a tall order for casual viewers at home. So a yellow field-line is digitally added to the broadcast to show where the team is aiming for. Instead of painting on the ground to make it look like an image, details are added to the viewing image so it looks like something is painted on the ground. Reverse situation, same maths.

For this to work, the computer system needs to know the exact direction every camera is facing and the exact shape of the playing surface: a football field is not perfectly flat but rather has a slight camber to help with water drainage. With this information the system can calculate precisely which pixels in every frame need to be turned yellow so it looks like there is a line projected on the field, and so the line stays rock-solid even when the camera is panning and zooming (which also involves correcting for the distortions caused by the geometry of the camera lenses). When over 100 million people watch the Superbowl live every year, they have no idea they are also watching live projection-geometry.

Unsurprisingly advertisers have swooped in on the use of anamorphic art in sports. It started the classic way with ads printed directly on the sports ground. Given many of the TV cameras for sports broadcasts are in known, fixed positions, some of those ads were drawn to look like a popped-out, flat image on TV. The effect is that ads on the field look like they are inserted into the video feed. Sometimes, on wide shots of the whole field, you can see anamorphic ads viewed from the 'wrong' camera and they look strangely stretched out, completely ruining the illusion.

Then the technologies collided. Ads are now inserted digitally, like the first-down line, to appear in the TV feed as though they are painted on the ground. Some of these are even designed to look like on-field anamorphic ads. I need to say that again. Advertisements are now digitally inserted into the video feed to look like they were printed on the field in such a way that, from the camera's point of view, they would look like they were inserted into the feed.

I first saw this with my own eyes at a rugby game in

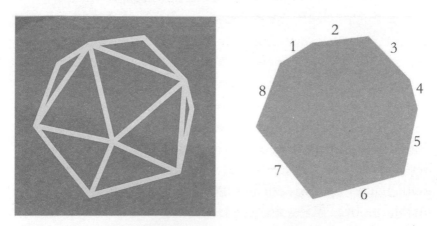

I regret throwing shade at this sign.

-0.998630

Two ads, one on the baseline just past the goal posts and one between the first two lines on the field, designed to look like an anamorphic image on the grass.

But, at exactly the same time in the game, you can see there is nothing but grass in those locations.

Australia. I noticed that the TV feed on one of the stadium big-screens showed ads which were not visible on the actual field. Then I figured out they were double anamorphic. I was unable to break down this geometric lineout to my rugby-obsessed friends, despite giving it my best try.

-0.997564

Projections also solve the problem of the Octagon Timber Flooring logo from the Introduction. I look at the logo and cannot help but picture it as a 3D icosahedron. In reality it is a 2D projection of an icosahedron, flattened so it can be printed on the sign. If you count the edges of the picture of an icosahedron there are exactly eight. The logo is an octagon after all! I just had to think outside the solid.

Impossible

Instead of using projections to make realistic art or insert convincing digital additions into video, we can also use the geometry of perception to make seemingly impossible sights.

The principles behind anamorphic art are used heavily in the creation of illusions. In 2023, the Best Illusion of the Year award was won by an anamorphic art illusion designed by mathematician and magician Matt Pritchard. This is a prize given by the Neural Correlate Society who support research into human perception and cognition. Finding ways our brains can be fooled gives us insight into how the human visual system functions.

Pritchard's entry was a cardboard model of a brick wall which a toy car is able to drive through. A real wall would have been even more impressive, but I suspect would have dramatically raised the budget. As it is, the video definitely makes you look twice as your brain struggles to reconcile what it thinks it is seeing with what it knows is impossible. This is a case where what your retinas are detecting is not best explained by the best explanation. The guesswork our visual system does all the time is suddenly exposed.

The same tricks can work with a physical object. It's possible to design very specifically shaped objects that trick our brains into making the wrong assumptions. These are shapes

-0.996195

One frame is not as impressive as the video of the car driving through a wall,
but you get the idea.

Looks like we've been had, folks.

that would never occur naturally and our brain makes a good guess at what it is seeing, which would be right were it not for the deliberately deceptive geometry. My favourite is the 'double arrow' designed by engineering-geometry specialist Kokichi Sugihara, who also invented the 'ambiguous cylinder'. They all take advantage of the fact that our eyes know the direction light must have come from, but not how far back along that path it originated. By carefully shaping an object, you can get human brains to confuse the far-away parts with the close-up parts.

You will need to reflect on these objects.

I decided to use the maths of projections to deal with a long-term issue I've had with street signs in the UK. The signs here for a football stadium use a depiction of a classic football (the soccer kind) which is normally made of a combination of pentagons and hexagons. If you flick back to the Archimedean solids in Chapter Six you'll see it there, listed as the truncated icosahedron. But do you know what you will not see anywhere in the Platonic, Archimedean, Johnson or indeed Anybody-at-all solids? A polyhedron made from only hexagons.

It is mathematically impossible to make a ball out of only hexagons. These British street signs are geometric fantasy.

-0.992546

Anyone who follows my work knows that I have campaigned to get these signs fixed, including going right to the UK government, and at every turn I have been unsuccessful. Now I have a new plan. Even though an all-hexagon ball is impossible, using projections we can make a ball which *looks* like it is made from only hexagons.

I asked hand-crafted ball maker Jon-Paul Wheatley if he could make me a football which looks exactly like the street sign . . . if viewed from one specific direction. Working with his partner Allison they were able to build a ball which, viewed directly from either the front or the back, looks like the street signs. Around the 'equator' between these two sides is a mash of the distorted pentagons and octagons required to patch the hexagons into a complete sphere.

All I have to do now is convince the English Premier League to switch to this new ball and instantly all of the street signs will be technically correct. But there is the problem. I took the ball up to Liverpool Football Club, one of the most prominent and successful football organizations in the

-0.990268

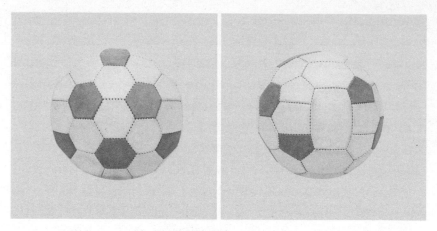

Behold, the impossi-ball!

world, and had a kick around with their sport-analytics team to see what they thought. They hated it. They said there is no way Premier League players would happily use a ball with an asymmetric pattern. I've started a petition to get the Premier League to use this ball all the same, but it seems for now the only projection illusions on a sports pitch are going to be adverts and first-down lines.

Thicker than a Ray of Light

Now we have an understanding of how boring, straight rays project onto flat surfaces, we can start to have some fun. The classic funhouse mirror uses a warped reflective surface to distort reflections to the point of hilarity. With mathematics we can trace every straight line that photons are travelling along, and fully understand the funhouse-mirror transformation. This also means we can flip the situation and design crazy images which only look normal when viewed in a distorting mirror.

I'm always one for taking maths to the masses, and every few years I embark on some ridiculous public maths project.

One time, it required getting a 2-metre tall, metal, mirror-finish pillar built. With a 1-metre diameter, it was a beast to move around (and fit through a doorway – ask me how I discovered that) but when set up it would offer a very distorted reflection of the environment around it. Looking back at you from the curved surface would be a very skinny version of yourself, and then an extreme wide-angle view of everything behind and a fairway to the side of you.

The project was to make images which appeared normal when viewed in the mirror-pillar but looked distorted in real life. The maths, of course, can be done in advance. When light hits a curved mirror the angle of incidence is still the same as the angle of reflection but now, like cycling up a Surrey hill, you only care about the gradient at one very specific location and direction. Which is our good friend tangent to the rescue. My team actually wrote some code for a website that made it possible to upload an image and convert it into distorted mirror-pillar form. School students and people at home made their own mini mirror pillars out of cardboard tubes and adhesive mirror-film.

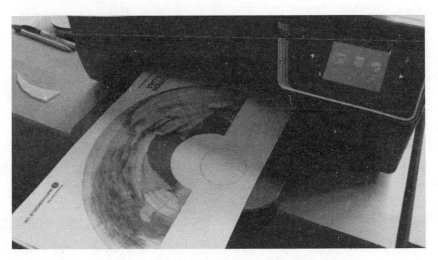

Print out a weird distorted photo of yourself.

-0.984808

But in a cylindrical mirror it looks great.

The massive physical pillar was an attention-grabbing device for the people and schools who do not already search out arty mathematics. My colleague Katie Steckles drove it around the country in a van and set it up in various libraries, museums and shopping centres. People would wander over for the funhouse mirror experience and Katie would hit them with the maths wonder. We generated a massive anamorphic grid that looks curved in real life but reflected in the pillar becomes a neat, square grid of straight lines. People could try their hand at drawing an image to add to the grid which would look normal in the pillar reflection. Someone crocheted a blanket with a pattern that only looked correct when viewed in the mirror pillar (it was a very big blanket). If you want to see it for yourself, the mirror pillar now lives with Maths City (on display in Leeds at the time of writing).

But all of this still assumes that light travels in straight lines. What if we get rid of even that idea? Can the maths still handle it? (I mean, of course it can, but let's build some narrative suspense all the same.)

-0.981627

Much of modern film VFX is based on 'ray tracing', which means following the mathematical paths of hypothetical rays of light. Like using a projector to beam out an image, it imagines beams of light coming out of the camera. Which is obviously not how visions or cameras actually work: in reality light comes off objects, bounces around the environment and ultimately may end up hitting an eyeball or camera sensor. Reversing the geometry to follow line-of-sight, rather than line-of-light, makes rendering a 3D environment more efficient. Only the photon paths which end up hitting the camera are calculated, and all other light can be ignored.

This is how the quadmesh environments (actually-triangle-meshes-but-whatever) we saw before turn a mathematical 3D object into a 2D frame of a film. The code starts with all of the pixels on the virtual camera sensor and beams out a 'ray' of light for each of them. As each ray moves around the computer-realized environment, it bounces off reflective surfaces until it eventually hits a solid object of some colour or texture. Each surface, colour, texture and substance the ray has hit, bounced off or passed through informs exactly what colour the originating pixel should be.

Which means the code doesn't care about the triangles outside the camera's field of view. I spoke to my VFX buddy Eugénie about this and she said that yes, any triangle which does not get hit by a ray is ignored completely. Moreover, things which are a long way from the camera, or only get hit by rays which have already reflected off something, can be rendered at a much lower resolution. So, while a large environment could run into billions or even trillions of triangles, thanks to ray-tracing they are never all used at once. And, very pleasingly, Eugénie calls the whole bundle of rays emitted from the camera the 'camera frustum'. Great use of 'frustum'!

Unlike painting, where the artist has to actually get a ruler out and line up the vanishing point, the ray tracing for VFX is all done automatically by the rendering software. It took some clever maths to code the software but, now it's done, surely the special effects artist does not need to understand the geometry behind the scenes. Which is true until you need to do anything slightly different. Like Eugénie did when she was working on Christopher Nolan's film *Interstellar*.

At the centre of the film is a giant black hole called Gargantua. Its incredible mass causes time dilation (like we saw with GPS satellites around the Earth), but another consequence of Einstein's General Relativity is that mass distorts the very shape of reality. A massive-enough black hole would cause spacetime to curve so much that light will noticeably no longer travel in straight lines. As this would very much be the case around Gargantua, Eugénie could not use an off-the-shelf ray-tracing program.

But because she has a very maths-heavy engineering degree, Eugénie was not worried about popping the bonnet and installing her own code. She was joined by Oliver James, Chief Scientist at Double Negative, as well as physicist Kip Thorne, who was their scientific advisor. A soon-to-be Nobel laureate, Kip would send pages of general relativity equations describing how photons should move in this exotic environment. Eugénie and Oliver would then turn those equations into working code that could drive a bespoke version of ray tracing.

I think the stunning visuals which resulted look so amazing partly because they are so physically accurate. Instead of a person guessing what a black hole would look like up close, the maths allowed the VFX team to be guided by reality itself. Which is not to say it wasn't an artistic endeavour. Cutting from the film's cinematography to a clean rendering of a hypothetical black hole would have been visually jarring. Like the first-down line being distorted to match the lenses on TV cameras, Eugénie had to distort, colour and flare the renders to make it look like they were shot on an IMAX camera floating in space. No visual effects shots exist in a vacuum (even ones set in a literal vacuum).

There were also times when they had to forgo some scientific accuracy because it made for a better story. Christopher Nolan's policy was always that the science could be as accurate as possible as long as it supported the plot, but if something had to give, it was going to be the science and not the narrative. In early test visuals, the team had included the calculations for a doppler effect shifting the colour of light coming from a spinning black hole. The right side of the black hole (as viewed from the camera) was moving away and so would be red-shifted and much dimmer, whereas the left side should be bluer and brighter.

The team at Double Negative tried simulating this but the shots of the spacecraft heading towards the black hole just didn't look right. To get the orbits right, as the spacecraft approaches the black hole it would be going towards the receding side on the right. But cinematically it would look very odd to audiences if the main character went towards the dingy side of the setting and not the bright side, which is film-language shorthand for the important side. So they turned off the doppler effect. Which made for a better film, but was no longer an exact recreation of reality. Classic art.

Ten
MAKING WAVES

L iving in the UK, I've learned to appreciate every possible minute of sunshine. I am at a greater angle of latitude here (51° from the equator) than where I grew up in Australia (a mere 32° from the equator). Because the Earth is on a tilt this means I get much more noticeable variation in the amount of daylight throughout the year. Which I kind of like, as it reminds me we're all clinging to a spinning atmosphere-covered rock hurtling through space around a giant star.

If the Earth was spinning in a perfectly upright position, with perfect posture (like people who use standing desks and have almost transcended their earthly form), we would have no such variation. Excluding anyone standing on the poles of the Earth, everyone would have exactly 12 hours of sunlight per day; you'd spend exactly as long on the far-side of the Earth as the Sun-side.

But that is not how the Earth rolls. It's slumped over on a 23.5° angle (like us mere mortals pouring ourselves into a desk chair every day). The direction of this tilt does not change as the Earth orbits the Sun, which means sometimes we are pointing towards the Sun (summer!) and sometimes

-0.965926

we are pointing away (brrr, winter). At the exact midpoint between summer and winter the tilt is at right angles to the direction of the Sun and the Earth's lean briefly no longer matters, giving us an equinox, with 12 hours of sunlight.

To see how the amount of tilt alters the length of our daytime, imagine if the Earth's axis was directly up: from the Sun's point of view you would move across in a straight line every day. Were the Earth transparent, your path would be you moving back and forwards on the same line. In the terrible case where the axis of the Earth pointed directly at the Sun, so at a 90° tilt, from the Sun's point of view you would move around in a circle. For all the tilts in between you would move around in an ellipse, the projection of a circle from the perspective of the Sun.

Trying to visualize all of these circles on a 3D, tilting Earth is a lot of pain, but after a while I was able to convince myself that the component of effective tilt towards the Sun was based on the sine of the angle between the Sun–Earth direction and the direction the tilt was pointing. Conveniently, a year-long orbit around the Sun is 365-ish days, very close to the 360 degrees in a circle, so the direction to the Sun changes 1° every day. With a bit of adjusting, I could get an equation for my amount of daylight: daytime = 4.34 × sin(day) + 12 hours.

Now. It is lovely to have a function involving sine which can take the day of the year and convert it into the number of hours of sunlight I can expect to hit my backyard, but it's not exactly a quick way to check the weather. Ideally it would be nice to have a simple weather chart I can look at. So I made a plot of the amount of light per day that I could get printed and hang in my house. Please note, this only works for latitude 51°N and if you live closer to the equator your wave will be flatter.

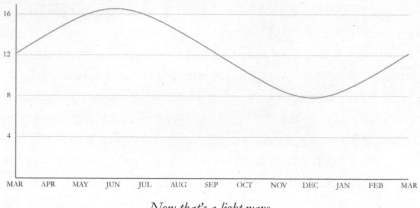

Now that's a light wave.

At this point a fair few readers will have shouted 'Finally!' Not because they are big fans of modern maths-art functional decor, although that may well also be true, but because this book finally contains a sine wave. If you plot the daylight time where I live (and most places on Earth) the resulting shape is something called a 'sine wave', and it is one of the most iconic shapes in mathematics.

I'll be honest: I actually plotted the daylight times before I knew what the graph was going to look like. After I loaded them into a spreadsheet and hit the 'plot' button, I immediately recognized the sine wave. It's a maths-celebrity graph. This is what led me on the quest to work through the geometry and convince myself why it was a sine wave.

On the surface it seems odd that the wave-shaped plot of a trigonometric function would be so prevalent, but it appears in so many places across mathematics, science and the world around us. For example, vegetables can make sine waves.

Grab some kind of cylindrical fruit or veg, like a courgette or carrot, and wrap a piece of paper around it. Instead of cutting it the normal straight-across way to give a circular end, try cutting it on an angle. When you unwrap the paper you will have a perfect sine wave.

−0.956305

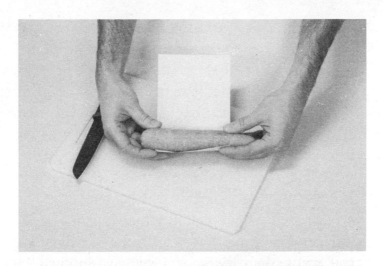

Carrots do help you see sine waves better.

With both the carrot and the daylight-plot, the secret is circles. Circles are how sine goes from being a property of an angle in a right-angle triangle to something you get after cutting a carrot. For many mathematicians, the concept of sine is more closely linked with circles than it is with triangles. When I claimed in the Introduction that Pythagoras was 'famously all about triangles, not circles' it hurt to type because I knew, deep down, that actually you cannot understand triangles and trigonometry without circles.

If you take a circle and 'pull it out into a spiral' (actually a helix), the projection of that helix from one direction is a sine wave and from a different direction is a cosine wave. (Sine and cosine waves are exactly the same shape, so 'sine wave' tends to get used to describe the shape in general.)

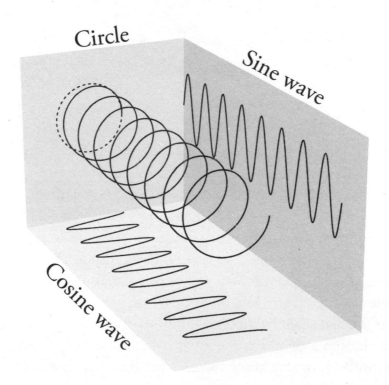

This is because every point on a circle is a right-angle tri-angle in disguise. The radius going from the centre of a circle to the outside is like the hypotenuse of a right-angle triangle and the other two sides appear when that point is connected to the horizontal and vertical axes going through the circle. If we consider the length of the hypotenuse to be a single unit, then the length of the other two sides will be sine and cosine. Meaning that the coordinates of any point on a circle is just the sine and cosine values of the angle in the middle. We have used a unit radius of 1, but in general the equation of a circle is $x^2 + y^2 = r^2$ for some radius 'r', which is just a bit of classic Pythagorean Theorem. Pythagoras is indeed all about circles.

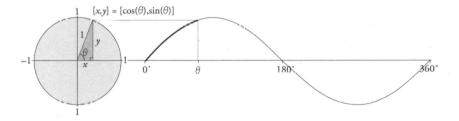

If we plot out all the values of sine for all the angles all the way around the circle from 0° to 360°, this gives us the sine wave. These are the three manifestations of what sine repre-sents: it's a function for angles in a triangle, a coordinate of a point on a circle, and a wave. In all cases the sine value starts at zero, increases gracefully to one, then descends to negative one, before rising back to zero and beginning the endless journey again. This also allows us to get sine values for any angle between 0° to 360°.

If we had stayed with our definition of sine as the ratio of two sides in a right-angle triangle we could never get a value for sin(90°), or any angle bigger than that, because at 90° it

-0.939693

would cease to be a triangle. Right before the triangle breaks, it makes sense that that value of sin(90°) is getting very close to 1 because the adjacent side and the hypotenuse are both absolutely massive, making their relative difference proportionally tiny: they are almost the same length and so the ratio between them is close to 1.

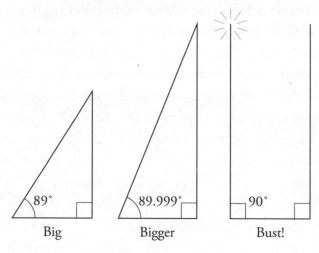

Scalene is not to scale.

At the moment the triangle breaks, the value of sin(90°) = 1 and now there is no opposite side and so cos(90°) = 0. Mathematically, it is pleasing to follow the logic of these values when the triangle hits its limit, but we don't actually need to do so. For any value between 0° and 360° you have two options to find the sine of cosine values: either check the coordinates for that point on a circle, or the heights on a sine wave. (A cosine wave is exactly the same shape as a sine wave, just moved to the side.)

This is handy for all the many applications of trigonometric functions we have already seen in this book, but there are two things about the values of sine and cosine which catch a few people off guard. For a start, some of the values are negative. Sine will give a negative value for anything between 180° and 270°, and cosine is negative from 90° through 270°. You can see this where the sine wave goes below the axis.

Secondly, there are multiple angles which give the same sine (or cosine) value. For example, both 70° and 110° have exactly the same sine value of 0.9397 which can cause some confusion. If I told you the sine of some angle was 0.7071, you would not know for certain if that angle was 45° or 315°. Or even 405°. In fact, there are infinitely many angles which correspond to any given sine or cosine value. This is partly because there are always two points on a circle for any value on one of the axes, and partly because you can keep rotating around a circle as many times as you want, going through the same points over and over.

We can see both of these on my daylight sine wave. The amount of sunlight is dependent on the sine of the angle representing how far through the orbit the Earth has travelled. This makes the orbit like a giant unit circle and the Earth a point racing around it, with the sine value the amount of sunlight.

I do not actually get negative sunlight, of course; rather, the values go above and below an average. For my daylight data, the average is 12 hours of daylight (actually 12 hours and 12 minutes if you factor in refraction and the size of the Sun) and so the equation of daytime = 4.34 × sin(day) + 12 hours means that sometimes sine is positive, and giving more daylight than average, and sometimes it is negative, taking daylight away. The value of 4.34 is to scale up the normal sine range of −1 to 1 to match the variability of sunlight at the latitude where I live.

For any particular amount of sunlight, I will get two such days in the year. For example, 10 hours on one day going into

-0.927184

summer and one day on the way back out in autumn. Then the whole pattern of daylight values repeats, year after year, as the Earth laps the Sun.

Part of the power of sine waves is their adaptability. You can move them, scale them and shift them. So far we have moved the sine wave up, and scaled it to be bigger. We can also shift a sine wave from side to side. Which I have actually done my daylight diagram. I wanted the 'beginning' of the wave on the equinox, which happens on 21 March, eighty days into the year. So technically my equation is daytime $= 4.34 \times \sin(\text{day} - 80) + 12$ hours.

Fun fact: up until 1752 England celebrated the new year on 25 March, which is absolutely close enough. Starting the year on the spring equinox makes so much more sense to me than in the middle of winter. But no, they had to go and move it to January. We could have had a perfectly aligned sine wave of a year if the 'calendar year' had been left where it was.

The British parts of North America also started the year on 25 March until 1752, only twenty-three years before the American Revolutionary War started in 1775. There are a lot of things which were common between the USA and the UK before the 'big split', which have since changed in the UK but remain fossilized in place in the USA, like a massive time capsule. The metric system is one example. If the revolution had been a mere quarter-century sooner, or the calendar change later, patriots in America would be defending the March 25 new year as strongly as they insist on using imperial units.

But What About Other Waves?

Sine waves can be produced from geometric situations involving circles but there are loads of other waves out there.

-0.920505

Sound waves, light waves, ocean waves. Are they also sine waves or not? It would be handy if they were because we could use all the mathematical tools from trigonometry and apply them to these other areas. We'll start with sound waves and find out what they can tell us.

We're going to start with one aspect of a single noise-making device: a guitar string. Why does it make noise? Because it vibrates. Why does it vibrate? Because it has a restorative force proportional to its displacement. Which requires some unpacking. I'm not going to unpack why vibrating air molecules hitting an eardrum is perceived as sound, as that would enter the realms of biology and psychology, both well beyond my comfortable home shire.

A 'restorative force' is a corrective push. Imagine a child falling over. You can either push them back upright or you can help gravity and push them over. Neither option is without its problems, depending on whose child it is, but in one case you're providing a restorative force to return the child to its upright position and in the other case you're open to a lawsuit. Some situations in the real world come with their own restoring force for free. A swing in a children's playground is a good example: if it gets moved away from hanging directly down it will swing back into place.

The same applies to a guitar string: if you pluck it away from its natural resting place, the tension in the string will snap it back into position. Not only that, it's a proportional restoration force meaning the further out of whack it is pulled, the stronger the force trying to drive it back home. If you slightly move a string on a guitar you'll notice it moves quite easily. But the further it gets plucked, the harder it becomes to move.

When eventually you let go of the string it will indeed snap back into position but, because it has momentum, it will overshoot in the opposite direction until the restorative force once again drags it back to the centre, only to over-shoot again the other way. In a frictionless wonderland this would continue forever. Sadly it would involve playing the guitar in a vacuum, which is a famously bad environment for a music gig. In the real world, friction and air resistance mean the string will vibrate back and forth for a while before that movement eventually decays away to leave the string stationary once more. Some of that energy is lost as the sound waves that we hear, so it's no bad thing that the string experiences this resistance. Unless the person is playing 'Wonderwall'.

Now, what's to say the resulting sound wave from the vibrating string is specifically a sine wave? That wave could be any crazy shape! Rock 'n' roll don't conform to no rules. But yes, it does. Rock 'n' roll do conform to many rules, both musical and mathematical. (But not grammatical, it seems.)

The 'proportional' of 'proportional restorative force' means that the more you move something, the harder it will try to spring back, and that changes perfectly with distance. If you move something twice as far, the force trying to return it gets twice as big; three times as far, three times the force. To describe this type of vibration we need a function where the acceleration is proportional to the position, but in the opposite direction.

The punchline is going to be that a sine wave is a perfect match for any wave produced by a proportional restorative force. To understand why, we need to look at how fast a sine wave is changing. Sine values oscillate between 1 and −1, but not at a constant rate. Looking at a sine wave you can see it has flat tops and bottoms, implying a graceful change of direction.

In fact, our car-slowing friends in Lambeth Council used

the shape of a sine wave as another way to calm traffic. In 2020, they installed seven speed bumps in Brixton which were the shape of sine waves (just the one hump, not a repeating sine wave). They claimed that the shape meant a 'shallower initial rise, which is more comfortable and safer for cyclists' but that the overall height was enough to be 'uncomfortable to drive over at speed in a vehicle'.

But how can we describe the exact way that a sine wave starts smooth but then gets steeper before flattening out again at the top? Conveniently, through the wonders of mathematics, the speed a sine wave is changing exactly equals the cosine value at the same position. I've tried to show this on a diagram. It takes a moment to follow what is happening, because you are looking at how fast the sine line is going up or down compared to the cosine value, but at all times the speed of sine equals the value of cosine.

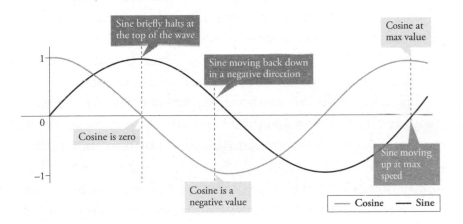

The reverse is almost true. Almost. The rate a cosine wave is changing is equal to the negative of the sine value at the same place. I am using 'speed' as a handy analogy from our normal human experiences because it is more intuitive to think of position, speed and acceleration. But mathematically we care about values, the rate those values are changing, and the rate those rates of change are changing. These are often called 'derivatives', but we don't need to get slowed down by that.

Speed of sin(θ) = cos(θ)
Speed of cos(θ) = −sin(θ)
⇨ Acceleration of sin(θ) = −sin(θ)

Combined, this means that the acceleration of a sine wave is equal to negative sine (and likewise for cosine). This is exactly the proportional restorative force required for sound waves. A pure sound wave is exactly a sine wave.

This is a hand-wavy way of proving that a vibrating guitar string is a sine wave, but I assure you it is backed up by a bunch more maths. Very few parts of this book are not a gateway to much more more maths, but I've had to draw a line somewhere. The point is, guitar strings produce sine waves. This is also true of piano keys, trombone slides and the almighty triangle (instrument, not shape). If a thing is vibrating and making waves, you're hearing sines.

Light waves are also sine waves. Light is often called 'electromagnetic' radiation because a photon is actually an oscillating pair of electric and magnetic waves. The physics behind this oscillation is a bit more complicated than a guitar string, but the result ends up being the same. Light waves are sine waves.

To avoid getting complacent, I included one red herring. Ocean waves are not sine waves. The forces that move water molecules in ocean waves are not proportional restorative forces. Rather than going up and down, the result of a bunch of fluid dynamics is that water molecules move around in a small circle as a wave goes by. This produces a wave shape called a 'trochoidal wave'.

Now that I think about it, music is also not pure waves (well, until my new genre of 'sine core' takes off) but rather a complex series of waves with depth and timbre. But don't worry. We can maths that as well.

Yeah, What About Other Waves?

I am not going to bury the lede: any wave, however complex and un-sine-like, can be represented with a combination of pure sines. Even the combined sound wave of a complete song. A song is made up of many sine waves, which combine to make something more intricate. But it is possible, if not easy, to do that in reverse.

Imagine you have a unique piece of piano music in two forms: the sheet music and a recording of the song being performed. Which would be worse to lose? If the recording was accidentally deleted, it would be possible to simply play and record the audio once more (assuming all recordings are created equal). But if the sheet music were lost, you would have to listen to the recording and try to decipher the chords back into the individual notes which were being played at the time – a long and tedious process.

What about the extreme version of this? Imagine that the recording is a much more complicated song with a wider range of sounds. Can a full recording of a song, with several different instruments and vocals, be reduced to sheet music playable on a piano? For a start, to even attempt something like this, we're going to need a fairly outrageous piano.

Theoretically it is absolutely possible. Mathematicians have proven that any audio, no matter how complicated, can be split into individual sine waves, aka notes. And when those notes are played together they will perfectly recreate the original sound. Any sound ever made is just a combination of sines.

-0.882948

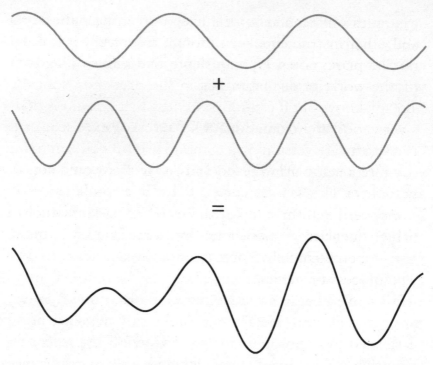

Sine waves: easy to add, hard to reverse.

Our slight practical problem, now, is that for a perfect recreation we would need a near-infinite piano with a lot of keys – as if a normal piano wasn't already hard enough to get up and down the stairs. As the number of keys, and therefore the number of frequencies, is reduced, we'll get a less accurate recreation of the original recording. Middle C is a frequency of 261.626 Hertz and next to it is a black key at 277.183 Hertz (C sharp), with nothing in between. There is no key, for instance, for anything around 270 Hertz. But with enough keys a piano could play a full song recording: guitar, drums, vocals and all.

Actual pianos can get close. The 'piano illusion' recreates full songs, lyrics and all, on a standard piano, but in such a way that no human could possibly play it. There are

-0.874620

examples online, and it feels like you can hear the vocals and other instruments even though technically it is only a mix of piano notes. Human suggestiveness is doing some of the work, as our brain fills in the blanks of songs we already know, but it cannot be denied that a piano is doing a very good approximation. (All the ones I have seen sadly use a software version of a computer. It's my dream to one day hire a self-playing piano and see if this would actually work.)

Converting from complex audio signals back into individual frequencies is not done by some kind of master transcriber, thankfully. There is a mathematical way to do it. The process of getting from a complex wave to an equivalent set of sine waves is called a 'Fourier transform'. Fourier analysis is like the greatest musical transcriber imaginable. It can listen to any sound and split it into its constituent frequencies. I think it is the most incredible bit of maths most people have never heard of.

It started with a French mathematician in the eighteenth century wondering how long he could hold a metal stick in a fire without burning his hand. You know, that classic game we have all played. But Joseph Fourier used the time it took for the metal to heat up to unbearable levels to think about some maths. As metal heats up, the atoms within it vibrate more and more energetically. Heat energy is able to travel through metal because these vibrating atoms cause waves of heat to move through the substance. These are effectively sine waves of heat moving up and down the metal rod, and calculating how long it would take for the heat to be unbearable required understanding how those waves interacted.

Eventually Fourier put the metal rod down, but he didn't let go of the idea. His new obsession culminated in a theory

-0.866025

that any function could be 'expanded' to a series of trigonometric functions which would sum back to give the original.
Fourier was doing this so the equations describing the
movement of heat could be turned into more manageable
trigonometric equations, but it ended up having a wide
range of other applications. Fourier didn't know it, but his
hot-stick obsession would one day change the world.

Fourier released *On the Propagation of Heat in Solid Bodies* in
1807 to only mild acclaim; according to one review, it 'still
leaves something to be desired on the score of generality and
even rigour'. It wasn't until 1878 that the work was even
translated into English. But despite this poor initial recep-
tion it has since become an all-time classic, like a kind of
mathematical *The Trig Lebowski*. It really did forever tie maths
and science together. Joseph Fourier now has his name
engraved on the Eiffel Tower in celebration of his achieve-
ments, which include the founding of Fourier analysis. Fun
fact: Fourier also discovered the greenhouse effect. How
topical (and, soon, tropical).

That is some very high praise indeed.

-0.857167

As waves became more important to our modern civilization, enabling us to do everything from understanding light waves and physics better to editing and transmitting audio signals, the theory of Fourier analysis hit the cold, practical reality of actually doing Fourier analysis. It was one thing to show that any wave could be split into sine waves, but that raised the question of how to actually do it. It turns out there are a lot of different methods, some more practical than others.

In a world before computers, physicist Albert Michelson built an analogue contraption at the University of Chicago which could do a Fourier transform using levers and springs. The Michelson Fourier Analyzer was available for purchase in 1904, and came in twenty and eighty sine-wave versions. The only known working version is owned by the mathematics department at the University of Illinois Urbana-Champaign. I went there on a Fourier pilgrimage, and they very kindly rolled the massive metal machine out of its glass cabinet and let me have a play.

The machine can do Fourier forwards and backwards. Forwards is easier. At the bottom are twenty different-sized

-0.848048

Back when sine waves were lovingly cranked by hand.

cogs, which move twenty levels up and down in a sine motion. These levers move metal rods that pull on a top bar via springs, so that the bar moves with the sum of all of the levers' individual motions. I was able to choose which levers I wanted to move and how much, each representing a sine wave, and when I cranked the handle, a pen attached to the top bar drew out a wave which was the sum of those sines.

It sounds complicated to follow and I assure you it is complicated to do. But even more complicated is working in reverse. Through a clever process of sliding metal bars around to match an input wave, the machine will draw out the arrangement required to recreate that wave from individual sines. As a largely untrained operator, I did not have much success. In the right, skilled hands this mechanical device could split any signal into a sum of up to twenty sine waves.

In a post-pre-computer world, there are a host of algorithms which can do this process to whatever level of precision is required, including a family of methods called 'fast Fourier transforms'. In all cases, it is a matter of comparing the input signal to a range of pure sine waves. Seeing how well each sine wave 'matches' the original (I'm hiding a lot of maths under the word 'matches') informs how much of that wave needs to go into the mix.

But how often do we really need to split a signal into its frequencies? I keep talking about how this underpins the modern world but, for that to be true, there needs to be a lot of situations that hinge on the decomposition of waves.

Spectr-O-Grama

It may seem frivolous, but Fourier analysis allows for an amazing graphics-equalizer display (those bouncy light things you sometimes see on stereos). The idea behind them is to show how loud a song is, not overall, but within different frequency bands. This is just low-resolution Fourier analysis.

These displays can be practical. When an audio engineer is getting ready for a concert or show they will do a bunch of 'EQ-ing', short for 'equalizing': reducing or boosting different frequency ranges. Music processing is of course a very common application of Fourier. It is possible to split a music signal into individual frequencies, move or remove some of them and then put the signal back together.

But the display of frequencies themselves can be useful, and not just a pretty visualization. A song is a constantly changing audio feed, which is why, if you want to play a tune on a piano, you need to keep changing which keys you're pressing. We can turn this idea into what is called a 'spectrogram', which is like a very dense form of sheet music. Time progresses from left to right, like in music, and the frequencies get higher as you go up the page. Unlike sheet music, which is largely binary (a note is either there or not), a spectrogram can use shading or colours to show how loud each frequency needs to be.

This turns a sound into an image in a way which gives more insight into the details of the noise. It is used extensively in science. I once spent a while in the Amazon rainforest with some researchers who are trying to track, count and document various species so the impact of human development can be detected and understood. I was there to make videos about the maths in their work. Some was fairly basic maths, like turning on a light in the forest at night and counting the different types of moth which appeared (and, I can confirm, the answer was a lot oh dear god they have flown down my shirt). But some of it was a bit more high-tech.

One of the researchers, Mark Bowler, is a primate specialist. He took me and my film crew out for a walk to show us how he and his team spot different groups of monkeys and follow their activity. It was not uncommon for researchers to spend a whole day tracking the same group of monkeys, not only to count exactly how many there were but also to log how far they were roaming. The problem was that this required a tedious day of being bitten by bugs while looking up. And occasionally dodging monkey waste, as my video producer Nicole and I discovered first hand. Apparently, not many people stick out the monkey-spotting job for long.

It would be convenient for everyone involved, both human and monkey primates, if this could be done automatically. The researchers plan to have automatic recorders in the forest which listen to the sounds monkeys make, and can correctly keep track of how many monkeys of each species are where on a constant basis. When I visited in 2023 they were still collecting recordings of monkey calls and manually identifying which species was which, with the goal of getting a sufficient set of training data that a machine-learning algorithm could take over.

I was fascinated to learn that when Mark is manually

recording which call was which monkey, he doesn't need to listen to each recording. He can look at a spectrogram and correctly identify what has made the noise. A bit like a pianist able to sight read, Mark can look at the changing frequencies over time and identify which monkey's call it is. Even we non-experts can discern some of the sonic details of the call in the Fourier spectrogram. A descending line is a lowering tone and a rising line, a rising tone. Lines which wiggle all over the place show how the tone changes in more complex ways.

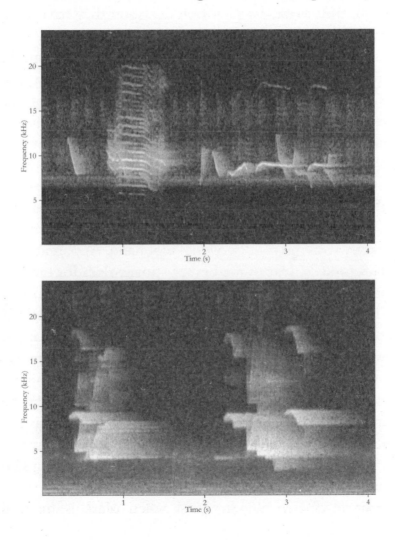

-0.809017

That first spectrogram is a Saddleback Tamarin call. Much of it is a single, varying tone but you can see one batch of parallel lines. This is a much more complex sound with several overlaid harmonics, which would be a very distinct call echoing through the rainforest. The second one is a Squirrel Monkey, and you can see some distinct lines high up on the plot. This is because the Squirrel Monkey has a very high-pitched aspect to its call. With the help of these plots, scientists are better able to track monkeys as part of their conservation work.

Spectrograms have also allowed humans to explore the universe on a scale almost beyond imagination. They allow us to detect phenomena called 'gravitational waves', for example, on a truly galactic scale. As is the scourge of GPS satellites, a large mass like the Earth distorts the shape of spacetime, which does weird things to the passing of time. But if the large mass also moves, then the distortion of spacetime can ripple out as waves in the fabric of reality itself. The first ever such gravity waves detected hit Earth on 14 September 2015, and we only noticed because of Fourier analysis.

This topic is dear to me because, many decades ago as an undergraduate at the University of Western Australia, I helped out with the early days of the Australian International Gravitational Observatory. Back then we could only dream of what a real gravity wave would be like and listen to simulated versions of what we might be likely to detect. Many years after I graduated and moved on to whatever this career is, it finally happened. I had also since married a research physicist, and so I heard a rumour at an astronomy festival that something big was about to be announced. Well, two big things: the gravity-wave signature of two coalescing black holes. The faint signal of two colliding black holes had ripped through the galaxy for over a billion years before reaching Earth and causing two different detectors, in Washington state and Louisiana, to fluctuate a near-imperceivable amount.

This required a huge amount of signal processing, which is partly why two detectors were needed, to confirm this was a real signal hidden in the noise (two detectors should have different noise but the same signal, making it easier to separate one from the other). Both detectors produced spectrographs which clearly showed the characteristic upward sweeping line of a black-hole collision chirp. Those two ticks represent colossal amounts of matter colliding with unimaginable energy at unfathomable distances from us. Here on Earth they caused spacetime to shuffle less than the width of an atom, but with phenomenal engineering, a lot of lasers and some clever maths we could detect that distinctive signal.

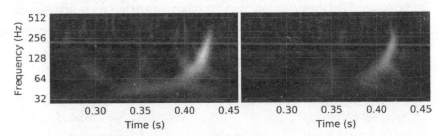

Signals detected in Washington and Louisiana, exactly offset by the speed of light distance between them.

Like the monkey calls, you can look at the gravity-wave spectrograms and imagine what the sound would be like. The upward rising signals are called a 'chirp'. The frequency of that chirp is below comfortable human hearing range, so it would actually be a very deep, seismic chirp, befitting some of the most dramatic, massive objects in the universe.

Never satisfied, scientists are now planning to put an interferometer in space to listen to much fainter gravity waves. Due to launch in 2035, the Laser Interferometer Space Antenna 'LISA' will consist of three spacecraft positioned 2,500,000 kilometres from each other. They will form the biggest triangle ever built by humans! The sides will be 200 times bigger than the Earth and be made of space lasers.

And for completeness here is a spectrogram of me saying my own name, 'Matt Parker'. It seems I manage to turn 'Matt' into a two-syllable word, which I find pretty funny. And the extra harmonics whenever I say the vowel 'a' are obvious, except I pronounce the 'a' in 'Matt' with a rising intonation whereas the 'a' in 'Parker' is on the way down. If I could memorize how to draw this spectrogram, then I would have a new, fun-but-tedious way to sine my name.

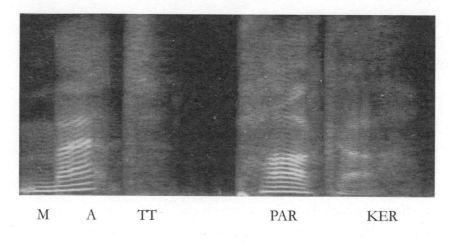

M   A   TT   PAR   KER

Let's Get Physical

Fourier analysis is not only useful for analysing audio signals; it can help us explore the very nature of matter. Crystallography is a method of probing the arrangement of atoms within a substance to determine the geometry of unimaginably small molecular arrangements. When early-career crystallographer Kathleen Lonsdale showed up in 1922 to start her master's project with esteemed scientist William Henry Bragg, there was a period of around three months during which her experimental equipment was still being organized. Bragg gave her a copy of *Mathematical Crystallography* to read in the meantime.

This book contained the state-of-the-art mathematics for crystallography, but if you search it for the word 'Fourier' (and I have) it returns exactly zero results. You'll see a lot of the words 'symmetry' and 'lattice', and there are trig functions for days. But no Fourier. A modern crystallographer would find this absurd; studying crystal structures without Fourier is now unthinkable. But this was a time after Fourier had done his work and before Lonsdale showed the world that it was the key to unlocking crystallography.

The then-cutting-edge book is a good snapshot of the mathematics in use before Lonsdale showed up. The book explains that if you are faced with a 3D lattice of atoms, there are 230 possible symmetry patterns they could have. These atomic arrangements are too small to be illuminated with normal light, but if they are blasted with short-wavelength x-rays, those x-rays will refract off the lattice and produce a pattern once they hit a screen. The pattern will not be an exact silhouette of the crystal lattice, but rather a series of shapes which have been produced by the various alignments within the lattice. The book instructed crystallographers to look for symmetry alignments in the projected pattern, and then work back to decipher what the crystal structure could possibly be.

This system did work. Scientists discovered that the structure of things like sodium chloride (table salt) was a cubic lattice with the atoms arranged in a perfectly regular grid, every sodium atom equally close to all the chlorine atoms near it. Chemists already knew that salt was equal amounts of sodium and chloride, but they assumed the sodium and chloride atoms were paired up in salt 'molecules'. William Henry Bragg discovered that there is no such thing as a molecule of salt because, for any given sodium atom, there was no one chloride atom it had a special relationship with. Salt was rather just a ratio of the two atom types. Bragg later recounted how he was begged by chemists to find a slight favouring of sodium to one of its chlorine neighbours 'in order that the molecular idea could be retained', but he refused.

Lonsdale had some early success with the old method, taking the photographic plates which had been exposed to x-rays diffracting off a crystal and looking at the patterns really hard. In 1929 she was the first person to discover that the heart of benzene molecules is a hexagon of carbon atoms. Specifically, a flat, regular hexagon – I would argue, the smallest physical regular hexagon in our universe. Then in 1931 she completely changed the game.

She was working on cracking the structure of a compound called hexachlorobenzene. It was proving to be much more complicated than sodium chloride, or even the hexagon-based benzene. Staring at the patterns produced by the x-rays, she saw that the way they were interacting with the lattice of atoms was performing a physical Fourier analysis. The regularity of the lattice was taking the place of a repeating wave frequency. She was able to take the pattern, run it through Fourier analysis, and out popped a map of where the atoms are inside the compound.

On 1 October 1931 the Royal Society published Lonsdale's paper, 'An X-ray Analysis of the Structure of Hexachlorobenzene, Using the Fourier Method'. That title really says it all: she was looking for the structure of hexachlorobenzene

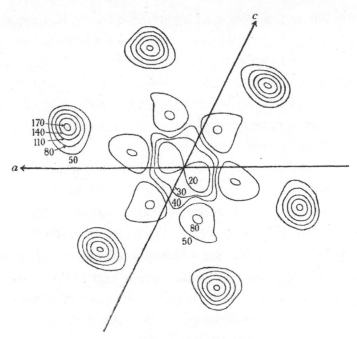

FIG. 7.—Projection of $C_6 Cl_6$ Molecule.

The electron-density contours that Lonsdale found using Fourier.
This is where the electrons be.

and used the Fourier method. This is the first ever use of Fourier analysis to determine the structure of a chemical compound.

Sadly it was a rather unfitting compound for such a great honour. In the 1940s hexachlorobenzene entered into use as a pesticide. It was banned by the Stockholm Convention on Persistent Organic Pollutants in the early 2000s, but not before this toxic and carcinogenic compound had caused much undue suffering. This now dominates any online information about the chemical and, despite its use being unrelated to knowing its geometric structure, has drowned out any accolades as the first molecule to be probed using Fourier analysis.

Lonsdale had a long and impressive scientific career. She was also a pacifist, a promoter of gaol reform and was named the joint-first female fellow of the Royal Society (in the embarrassingly recent year of 1945). Her early insight that Fourier analysis could reverse the interaction of x-ray waves to probe the very nature of matter around us was an incredible step forward for humankind.

Many years later, when Rosalind Franklin was taking x-ray diffraction images of DNA which led to cracking the double-helix structure, she used the same Fourier techniques pioneered by Kathleen Lonsdale to reverse engineer the images. Much focus is rightfully placed on Franklin, but I think we should also appreciate the advancements made by Lonsdale a few decades prior. Thanks to the incredible insight of these two scientists, we now understand the nature of DNA which makes all life possible.

And it is extra-pleasing that the sine waves needed to understand DNA are exactly the same as the shape a helix makes when viewed from the side.

Conclusion

In conclusion, everything is triangles. It's no wonder the Illuminati symbol is an all-seeing eye inside a triangle. Any object can be represented and simulated by a triangle mesh (or lattice), and any signal can be built up from sine waves. There is nothing triangles cannot do.

I would like to apologize to anyone whose favourite bit of triangle-related mathematics didn't make it into this book. Me as well! I'm sure you kept waiting for it to pop up, and now you're reading the Conclusion and it never made an appearance. I'm sorry. There was simply too much candidate maths which could have gone in. I had to choose a single path through all the options and sadly the book is finite in length. Space-filling curves and fractals are one big topic I would have loved to cover, but there was not enough space.

As the book was nearing completion, one of my fact checkers (Adam Atkinson, who has been through all my books with a fine fact-comb) asked if I was going to include prosthaphaeresis in the section on trigonometry. I was certainly tempted! Prosthaphaeresis was a technique used between about 1590 and 1614. If you wanted to multiply two large, cumbersome numbers together, you would find angles where the cosine of each equalled one of your numbers (scaled between 0 and 1) and then use a trig identity to turn multiplication into addition.

But adding that to the book would require covering a lot of additional ground just to explain how it worked. In the end, it went on the 'topics I love but cannot fit in' pile. I don't know what the angle of repose of that pile would be.

I'm also sorry for the dead ends. Mathematics is so interconnected that it makes extracting a single, logical path impossible without tearing a few bits off. In the second chapter, I mentioned that the wake behind a duck is 39°, which actually holds for any object moving in sufficiently deep water. This, I wrote, 'tells us something about the way waves move in water', and I then moved on, never to swim back to that concept.

Well, now we have covered waves and how they interact, combine and cancel out, so technically we could tackle the hydro-ducknamics, but we're out of room. You'll just have to take my word for it that the way the waves interfere and constructively add in one, single, V-shaped maximum is exactly the same mathematics that explains how light leaving a water droplet adds to give a single, maximum angle of brightness. Duck wakes and rainbows are both the result of the same bit of maths. Humans have been looking at rainbows and watching ducks since prehistory, and it's only within the last few centuries that we've had the mathematical knowledge to understand they are one and the same.

I followed the path of triangles → geometry → trigonometry → sine waves because I really wanted to get to waves and Fourier analysis. I hope you were sufficiently convinced that we cannot understand waves mathematically if we don't first understand triangles, thus justifying everything's position in this book.

I don't think it would take much arguing to persuade anyone that an understanding of waves makes our modern, digitally connected world possible. Signals fly at the speed of light between mobile phones and communication towers, as well as down fibre optic cables between continents. All impossible if we didn't have the knowledge of Fourier to disentangle multiple, combined sine waves.

Understanding waves can also help in your normal life. The distinctive sound of the band Queen, one of the best-selling music acts of all time, was partly down to Brian May's guitar which he built himself. Six years before Queen, when he was still a teenager, Brian built the guitar, with the help of his dad, from partly improvised components including wood salvaged from an old fireplace and motorbike valve springs. Already a bit of a science nerd, Brian wired the three electric pick-ups himself, in an unusual way, with extra switches that allowed him to flip the wiring on any of them. This would change the 'phase' of the sound-wave signal coming from that pick-up, effectively turning the sound wave upside down. Highs become lows and lows become highs.

This flipping changes how the signals from the pick-ups interact, and alters which parts of the sound constructively add and which cancel out. Different switch settings would give different feelings to the resulting sound, and Brian would change them depending on the song. For example, if one of the 'middle' and 'neck' pick-ups are flipped so they are out of phase with each other – something impossible to do on any normal guitar – it gives the iconic sound of the guitar solo in 'Bohemian Rhapsody'. That song would not have been possible if a young Brian May hadn't understood how waves interact mathematically.

I was so fascinated with how Brian May's guitar, named the Red Special, messed with sine waves that (despite my established amusicality) I wanted to see if I could simulate it. I worked with maths and music expert Ben Sparks to make an interacting waveform display which came with three digital switches to demonstrate exactly what Brian's wiring was doing to the audio signal. It all ran on an iPad. Brian May

-0.694658

is famously an astronomy nerd (now with a PhD in the subject) and attends the astronomy festival my wife helps organize. In the same year that the gravity-wave rumours were going around, I was talking to Brian and took the opportunity to show the simulation to him. He was very impressed, describing it as 'absolutely lovely', and I suspect it may be the most mathematical piece of fan art he's ever received.

I don't know if that counts as maths being useful in someone's career or doing geometry because it is fun. It certainly helped Brian May's career and produced some of the most recognizable songs of the past century, but he also does nerdy things as a hobby. I certainly simulated the waves interacting purely because it was fun, but now I'm writing about it in a book which I think technically counts as work (at least, it does according to my accountant). It all overlaps and interacts constructively.

The final wave kicker comes from quantum mechanics. We've already had a look at the sibling area of modern physics, General Relativity, which describes how things behave when they are very big or moving very fast. Quantum mechanics deals with the opposite end of the spectrum: the very, very small. If you zoom in far enough, the intuitive reality around us fades away and is replaced by pure mathematics. We live in a universe made of maths. What we think of as solid matter is actually wave functions.

You are literally made of sine waves. You are built of triangles. Reality is triangles.

Triangles are everything and everything is triangles.

And now we have that out of the way: don't worry, while I was writing this book they fixed the biscuit packaging.

CHEESY
THINS

POPPY SEED
CRACKERS

ACKNOWLEDGEMENTS:
a triangle of thanks

Agents, editors and producers:
Courtney Young, Laura Stickney, Nicole Jacobus, PJ Mark, Richard Atkinson, Sarah Cooper, Will Francis and everyone at Penguin Random House, Janklow & Nesbit and the whole Stand-up Maths machine.

People who worked on the book:
Images and photos by Alex Genn-Bash, Jennie Vallis, Sam Hartburn, Simon Kallas, Truman Hanks. Fact checking by Adam Atkinson, Charlie Turner, Colin Beveridge with additional mistakes spotted by Jack Craig, Jean-Philippe Belmont. And these people who helped above all reasonableness: Eugénie von Tunzelmann, Laura Taalman, Paul Shepard.

Those who shared their time, expertise and ideas:
Adam Savage, Alex James, Allison Wheatley, Andrew Pontzen, Ben Sparks, Beth Crane, Bill Gosper, Bill Hammack, Bill Hedges, Chaim Goodman-Strauss, Chris, Chris Fewster, Clara Grima, Darren Morgan, David Grace, David McCabe, David Smith, Em Bell, Flic Luxmoore, Garrett Ryan, Geoff Lindsey, Grant Sanderson, Hannah Fry, Hanyu Alice Zhang, Helen Arney, Henry Segerman, James Bull, James Grime, Jennifer Barretta, Jon Harvey, Jon-Paul Wheatley, Katie Steckles, Kevin Armstrong, Lucie Green, Maddie Moate, Maggi Grace, Mark Bowler, Matt Pritchard, Nick Harris,

Oliver Kirkpatrick, Phil Green, Randy Linden, Rob Eastaway, Robert Austin, Robin Houston, Rollie Williams, Sabina Raducan, Sara Morawetz, Seb Lee-Delisle, Steve Mould, Tim Chartier, Tim Waskett, Timon Gutleb, Trent Burton, Vincent Gallo and the many other people who have helped but I accidentally neglected to list here.

Picture Credits

In addition to the photos from my personal collection, the professional-looking custom photos in this book were taken by Simon Kallas, Alex Genn-Bash and Truman Hanks.

Illustrations made by Jennie Vallis and Sam Hartburn, other terrible plots by me.

Thanks to everyone who let me use photos they had taken: Colin Leonhardt for the double rainbow, Enric Florit for the UFO bar in Barcelona, Laura Taalman for mathematical bracelets and scutoids, Vincent Austin Gallo for bees doing things (but not maths), Timon Gutleb's wedding cake, Matt Pritchard's illusions, Kristen Lomasney for taking a shot of me standing next to Adam Savage, and Phil McIver for their great photo of Fourier's name on the Eiffel Tower.

Stock images from Shutterstock, Wikimedia Commons, Alamy and Pixabay. All terrible photoshopping was done by me.

The Moon in front of the Earth, the parallax images of Proxima Centauri and Wolf 359, Saturn wearing a hexagon hat, JW Space Telescope photos and deep field images are all thanks to NASA's very generous public domain policy.

Papyrus photos taken by me but under the watchful and permitting eye of the British Museum.

Photos of me on the MotoGP bike are photo credit Bonnie Lane. Image use is courtesy of Two Wheels for Life and kind arrangement with The Cosmic Shambles Network and Dorna Sports, S.L.

The 3D UFO model is from TurboSquid and Eugenie von Tunzelmann helped render both it, and my low-triangle face. Paul Shepherd provided some 3D graphics of engineering things. Monkey-call spectrograms are thanks to Mark Bowler spending endless hours in the rainforest.

-0.615661

Asteroid impact images from 'After DART: Using the First Full-scale Test of a Kinetic Impactor to Inform a Future Planetary Defense Mission' by Thomas Statler, Sabina Raducan et al (The Planetary Science Journal, 2022) and 'Physical properties of asteroid Dimorphos as derived from the DART impact', by Sabina Raducan et al (*Nature*, 2024).

The model of Donald Grace's shape was from 'Search for the Largest Polyhedra' by Donald Grace (Mathematics of Computation, the American Mathematical Society, 1962).

Perspective lines drawn on a painting from 'Perspective as a geometric tool that launched the Renaissance' by Christopher Tyler (Proceedings of SPIE - The International Society for Optical Engineering, 2000).

Gravity wave spectrogram from 'Observation of Gravitational Waves from a Binary Black Hole Merger', by B. P. Abbott et al. (LIGO Scientific Collaboration and Virgo Collaboration) published in Physical Review Letters 2016.

Atom plot from 'An X-ray analysis of the structure of hexachlorobenzene, using the Fourier method', by Kathleen Lonsdale (Proceedings of the Royal Society A, 1931).

Index

This is a three-point index. Every entry is for a sequence of three words which appears somewhere in the book. You also get three bits of information: the three words, the distance from the bottom left corner of the page (in units of page width) and the angle from the corner to the start of the words (degrees counter-clockwise).

-0.587785

anything other than: 0.898794 (d = 0.611, θ = 50.4°), -0.939693
(d = 0.794, θ = 75.0°)
aperiodic monotile, an: 0.544639 to 0.544639, 0.500000 (d = 1.176, θ = 64.7°)
aperiodic monotile, chiral: 0.469472 to 0.469472
Archimedean solids, The: 0.241922 to 0.190809, -0.992546 (d = 0.833,
θ = 34.4°)
Archimedean solids, thirteen: 0.156434 (d = 1.189, θ = 53.2°), 0.017452
(d = 0.638, θ = 65.7°)
Area Base Height: 0.898794 (d = 0.595, θ = 58.8°)
around the Earth: -0.500000 (d = 0.903, θ = 50.1°), -0.743145 (d = 1.185,
θ = 53.5°), -0.809017 (d = 1.256, θ = 49.3°), -0.906308 (d = 0.826,
θ = 28.7°), -0.974370 (d = 1.139, θ = 69.0°)
asteroid impacts, an: 0.798636 (d = 1.309, θ = 60.3°), 0.819152
(d = 1.148, θ = 82.2°), 0.838671 (d = 0.613, θ = 71.6°)
Asteroid Redirection Test: 0.829038 (d = 1.490, θ = 65.9°)

back and forth: 0.681998 (d = 0.694, θ = 42.7°), 0.939693 (d = 1.059,
θ = 56.0°), -0.906308 (d = 0.862, θ = 19.3°), -0.906308 (d = 1.319,
θ = 66.7°)
balloon was, the: 0.156434 to 0.173648
base-height theorem, the: 0.933580 to 0.933580
baseball diamond, a: -0.573576 to -0.587785
basketball hoop, the: 0.945519 (d = 0.401, θ = 45.3°), -0.390731
(d = 1.221, θ = 82.7°)
Beach Rescue At: -0.438371 (d = 1.283, 0 = 70.8°)
better version, a: -0.034899 (d = 0.860, θ = 24.7°), -0.882948 (d = 0.807,
θ = 59.1°)
big circle, a: 0.000000 to 0.000000
bike ground angle: 0.996195 to 0.997564
Bilinski dodecahedron, the: 0.139173 to 0.121869
billion light years: 0.438371 (d = 1.230, θ = 53.6°)
bit more complicated: -0.422618 (d = 1.258, θ = 46.6°), -0.945519
(d = 0.683, θ = 58.4°), -0.891007 (d = 0.659, θ = 55.8°)
bit surprising, a: 0.104528 (d = 1.151, θ = 81.3°), -0.656059 (d = 0.809,
θ = 26.4°)
black hole, a: 0.406737 (d = 1.116, θ = 72.3°), 0.978148 (d = 1.048,
θ = 78.3°), -0.974370 (d = 0.573, θ = 38.8°)

Bohemian Rhapsody That: -0.694658 (d = 0.624, θ = 68.9°)

brains, and our: -0.965926 (d = 0.739, θ = 35.9°), -0.994522 (d = 0.671, θ = 36.3°)

Brand New Shape: 0.292372 (d = 1.287, θ = 70.3°)

Brian May's guitar: -0.694658 to -0.694658

brick wall, a: -0.275637 (d = 0.572, θ = 62.7°), -0.996195 (d = 0.982, θ = 41.9°)

British Museum Because: 0.309017 (d = 0.917, θ = 56.8°)

British Museum roof: 0.994522 to 0.994522

British Museum, the: 0.309017 to 0.309017, 0.994522 to 0.992546

Burroughs Donald Grace: -0.629320 (d = 0.580, θ = 28.4°)

Burroughs was, The: -0.121869 to -0.139173

calculate how big: -0.500000 to -0.515038

camera, from the: 0.207912 (d = 0.611, θ = 40.5°), 0.731354 (d = 1.009, θ = 70.7°), -0.974370 (d = 1.283, θ = 66.3°), -0.998630 (d = 0.852, θ = 64.2°), 0.978148 to -0.978148, -0.970296 (d = 0.999, θ = 66.3°)

Catalan solids, The: 0.207912 (d = 0.892, θ = 44.0°), 0.156434 (d = 0.900, θ = 69.2°)

Chambers's Shorter Six-Figure: -0.241922 (d = 1.378, θ = 80.6°)

Chris Fewster, Professor: 0.156434 (d = 1.078, θ = 51.3°)

Christmas tree, a: -0.390731 (d = 0.614, θ = 55.8°), -0.422618 (d = 0.834, 0 = 37.5°)

Christmas tree, my: -0.390731 to -0.390731, -0.669131 (d = 1.290, θ = 55.5°)

circle, a small: 0.000000 to 0.000000, -0.891007 (d = 0.934, θ = 20.2°)

clever maths, some: -0.974370 (d = 1.332, θ = 83.3°), -0.788011 (d = 0.952, θ = 74.0°)

Closer Matt, Because: -0.984808 (d = 1.600, θ = 63.6°)

computer, The original: -0.601815 (d = 0.709, θ = 67.7°), -0.669131 (d = 0.457, θ = 38.7°)

contacted my friend: 0.951057 (d = 1.151, θ = 67.5°), 0.987688 (d = 1.012, θ = 77.9°)

convince myself that: 0.422618 (d = 1.547, θ = 61.5°), -0.961262 (d = 1.209, θ = 46.6°)

Cosine Rule, The: -0.559193 to -0.573576, -0.766044 (d = 1.008, θ = 49.5°)

-0.559193

cosine value, or: -0.615661 (d = 1.664, θ = 59.5°), -0.927184 to -0.927184

cosine wave, a: -0.945519 (d = 1.249, θ = 55.2°), -0.933580 (d = 0.494, θ = 53.4°), -0.898794 (d = 0.836, θ = 28.3°)

Cosmic Films Studio: -0.139173 to -0.139173

Cosmic Web, the: 0.390731 to 0.406737, 0.438371 (d = 1.274, θ = 53.6°), 0.573576 (d = 0.791, θ = 75.8°), 0.933580 (d = 0.607, θ = 28.1°)

DART spacecraft, the: 0.829038 to 0.829038, 0.891007 (d = 0.860, θ = 53.7°)

daylight time, the: -0.956305 to -0.956305

daylight, plot the: -0.956305 to -0.956305

Degree Confluence project: -0.891007 (d = 1.217, θ = 74.7°), -0.920505 (d = 1.667, θ = 59.3°)

digital camera, a: 0.898794 (d = 0.621, θ = 69.5°), -0.961262 (d = 0.665, θ = 21.9°)

disdyakis triacontahedron, a: 0.999391 (d = 1.102, θ = 68.6°), 0.974370 (d = 1.337, θ = 81.7°)

disdyakis triacontahedron, the: 0.999391 to 0.998630, 0.207912 (d = 0.808, θ = 20.7°)

distance between two: 0.978148 (d = 1.249, θ = 65.6°), -0.838671 (d = 1.399, θ = 60.0°)

distance between, the: 0.978148 (d = 1.229, θ = 67.8°), -0.743145 (d = 1.058, θ = 69.2°), -0.819152 to -0.819152, -0.838671 to -0.838671, -0.913545 (d = 0.584, θ = 60.5°)

distance from us: 0.469472 (d = 1.320, θ = 52.3°), -0.992546 (d = 1.179, θ = 64.4°), -0.788011 (d = 1.178, θ = 60.6°)

distance, Take the: 0.939693 to 0.939693

divided by opposite: -0.309017 to -0.309017

dodecahedron, a regular: 0.052336 (d = 1.263, θ = 81.2°), 0.017452 to 0.017452

dodecahedron, the regular: 0.642788 to 0.629320, 0.275637 (d = 0.942, θ = 35.3°)

Dodecahedron, the Rhombic: 0.642788 to 0.615661, 0.587785 to 0.573576, 0.207912 (d = 0.667, θ = 28.9°), 0.139173 to 0.139173

Donald W Grace: -0.069756 (d = 0.514, θ = 33.8°)

Double Negative, at: -0.974370 to -0.970296

double rainbow, a: 0.777146 to 0.777146

duck, behind a: 0.601815 (d = 1.350, θ = 78.5°), -0.707107 (d = 1.363, θ = 66.3°)

earliest known human: 0.438371 to 0.438371

Earth, from the: 0.190809 (d = 0.567, θ = 74.0°), 0.515038 (d = 0.975, θ = 61.9°)

Earth's radius, the: -0.694658 to -0.707107, -0.766044 to -0.766044, -0.838671 to -0.838671

Earth's, and the: -0.719340 (d = 1.334, θ = 52.8°), -0.766044 (d = 0.972, θ = 48.7°), -0.961262 (d = 1.414, θ = 74.0°), -0.927184 (d = 1.104, θ = 44.0°)

Einstein's General Relativity: -0.974370 (d = 1.064, θ = 74.6°)

elongated square gyrobicupola: 0.156434 to 0.156434

engineering mate, my: 0.981627 to 0.981627

engineers loved triangles: 0.981627 to 0.981627

English Premier League: -0.990268 (d = 0.738, θ = 32.5°)

equator, from the: 0.529919 (d = 0.698, θ = 42.4°), -0.819152 (d = 0.713, θ = 49.9°), -0.920505 (d = 1.464, θ = 78.4°), -0.965926 to -0.965926

equilateral pentagon dodecahedron: 0.104528 to 0.104528

equilateral triangles, and: 0.731354 to 0.719340

Eugénie von Tunzelmann: 0.974370 (d = 0.960, θ = 30.6°), -0.642788 (d = 0.740, θ = 80.4°)

European Space Agency: 0.891007 (d = 1.045, θ = 71.1°)

eye, the human: 0.992546 (d = 1.216, θ = 48.8°), -0.777146 (d = 0.898, θ = 18.5°)

Family Guy, and: 0.017452 (d = 0.949, θ = 37.8°), 0.927184 (d = 0.625, θ = 54.4°)

farmers had, the: 0.156434 to 0.156434

feet and inches: 0.951057 (d = 1.463, θ = 68.3°), 0.961262 (d = 1.493, θ = 65.6°)

film and TV: 0.970296 (d = 1.271, θ = 52.5°), 0.939693 (d = 1.218, θ = 84.2°)

Finite element analysis: 0.956305 to 0.956305, -0.190809 (d = 0.787, θ = 27.2°)

first-down line, the: -0.999391 to -0.998630, -0.970296 (d = 1.409, θ = 82.4°)

first-down, like the: -0.998630 (d = 0.983, θ = 75.6°), -0.970296 (d = 1.402, θ = 84.9°)

Fourier analysis, Physical: -0.777146 (d = 0.679, θ = 39.8°), -0.754710 (d = 0.800, θ = 42.9°)

-0.529919

Fourier Analyzer, Michelson: -0.848048 (d = 1.004, θ = 80.0°)
Fourier Method That: -0.754710 (d = 0.509, θ = 34.1°)
Fourier Method, the: -0.754710 to -0.743145
Fourier transform, a: -0.866025 (d = 1.095, θ = 60.4°), -0.848048
  (d = 1.073, θ = 73.0°)
Fourier, But Joseph: -0.866025 (d = 0.790, θ = 56.2°)
Fourier, The Michelson: -0.848048 (d = 0.997, θ = 82.9°)
Future Planetary Defense: 0.848048 (d = 1.434, θ = 67.3°)

Gamma-Ray Burst Explorer: 0.422618 (d = 1.274, θ = 63.3°)
gamma-ray burst, a: 0.406737 to 0.406737
gaps, fills the: 0.642788 (d = 0.721, θ = 65.4°), 0.104528 (d = 0.941,
  θ = 68.9°)
General Relativity These: -0.945519 (d = 0.852, θ = 46.6°)
geometry and trigonometry: 0.017452 to 0.034899, 0.069756 (d = 0.786,
  θ = 47.8°), 0.104528 (d = 0.949, θ = 43.7°)
Giant Ring, the: 0.438371 (d = 0.764, θ = 72.9°), 0.544639 (d = 0.846,
  θ = 29.4°)
Giant's Causeway, the: 0.838671 (d = 0.755, θ = 75.1°), 0.798636
  (d = 1.544, θ = 68.2°)
glass dome, a: 0.999848 to 0.999848
Global Positioning System: -0.882948 (d = 1.431, θ = 77.4°)
GPS device, a: -0.891007 (d = 1.427, θ = 72.3°), -0.927184 (d = 0.453,
  θ = 52.6°)
gradient, per cent: -0.258819 to -0.275637, -0.438371 (d = 1.395,
  θ = 56.0°)
Greenwich observatory, the: -0.906308 to -0.906308
Grid North, Because: -0.920505 (d = 1.294, θ = 55.3°)
guitar string, a: -0.913545 to -0.913545, -0.891007 (d = 0.887, θ = 37.9°)
Guns N Roses: -0.996195 (d = 0.392, θ = 66.6°)

Hannah, my friend: 0.224951 (d = 0.599, θ = 65.4°), -0.754710 (d = 1.163,
  θ = 61.9°)
Hat family, the: 0.484810 to 0.469472
Hat tiling, The: 0.515038 (d = 0.555, θ = 64.6°), 0.469472 (d = 0.759,
  θ = 54.6°), 0.052336 (d = 0.978, θ = 69.3°)
Heron's Formula has: 0.857167 to 0.848048

hexagon, a regular: 0.838671 (d = 1.099, θ = 32.0°), 0.719340
    (d = 0.836, θ = 28.3°), 0.573576 (d = 1.277, θ = 43.4°), -0.052336
    (d = 0.736, θ = 26.0°)
hexagon, the regular: 0.766044 to 0.766044
Histories Book, The: 0.358368 (d = 1.316, θ = 71.6°)
horizon, at the: -0.754710 to -0.754710, -0.998630 (d = 1.481, θ = 52.1°)
hot-air balloon, the: 0.156434 (d = 1.133, θ = 76.2°), 0.207912 to
    0.207912, -0.857167 (d = 1.002, θ = 53.3°), -0.974370 to -0.978148
Hubble constant, the: 0.469472 (d = 1.185, θ = 55.0°), 0.544639
    (d = 0.691, θ = 40.9°)
Hubble Space Telescope: 0.829038 (d = 0.915, θ = 28.3°)

I'm a big: 0.087156 (d = 0.461, θ = 70.2°), 0.939693 (d = 1.194, θ = 67.1°)
identical vertices, and: 0.241922 (d = 0.989, θ = 44.5°), 0.207912
    (d = 1.493, θ = 69.3°), 0.139173 (d = 0.822, θ = 48.9°)
infinitely many, the: 0.694658 (d = 1.355, θ = 71.6°), -0.069756
    (d = 1.312, θ = 54.1°)
inside a triangle: 0.997564 (d = 0.592, θ = 37.4°), -0.719340
    (d = 1.304, θ = 64.3°)
Interstellar, the film: -0.945519 to -0.945519
isosceles triangle, an: 0.927184 (d = 0.867, θ = 28.6°), 0.965926 to 0.970296

J J Thomson: -0.069756 (d = 1.359, θ = 71.1°)
James Van Allen: 0.656059 (d = 1.541, θ = 68.5°)
Jet Propulsion Laboratory: -0.642788 (d = 1.069, θ = 43.1°)
Johnson solids, the: 0.190809 (d = 0.803, θ = 46.4°), 0.121869
    (d = 0.997, θ = 57.5°)
J W Space Telescope: 0.838671 to 0.829038

Ken Perlin's, Both: 0.656059 (d = 0.934, θ = 41.5°)
kites and rhombuses: 0.052336 (d = 0.914, θ = 61.6°), 0.017452
    (d = 1.006, θ = 40.7°)

Land's End Much: -0.857167 (d = 0.886, θ = 32.1°)
latitude and longitude: -0.798636 (d = 1.152, θ = 55.7°), -0.819152
    to -0.838671, -0.891007 to -0.898794, -0.913545 to -0.913545, -0.956305
    (d = 1.080, θ = 66.3°)

Nintendo Entertainment System: -0.656059 (d = 0.876, θ = 26.9°)

North Downs What: 0.052336 (d = 0.325, θ = 61.4°)

North True North: -0.927184 to -0.927184

North, from the: 0.573576 (d = 1.143, θ = 59.9°), -0.707107 (d = 0.323, θ = 50.3°), -0.798636 (d = 0.791, θ = 21.2°)

North, north Magnetic: -0.920505 to -0.927184

Octagon Timber Flooring: 0.069756 (d = 0.572, θ = 29.9°), -0.996195 (d = 1.603, θ = 63.4°)

octagon, not an: 0.069756 to 0.087156

oilfield worker, The: 0.034899 (d = 0.965, θ = 80.8°), -0.342020 (d = 1.054, θ = 46.7°)

One Annoying Step: 0.224951 (d = 1.348, θ = 72.7°)

One World Trade: 0.374607 to 0.374607

Ordnance Survey, the: -0.866025 (d = 1.200, θ = 55.5°), -0.920505 (d = 1.353, θ = 71.9°)

other way around: 0.777146 (d = 1.068, θ = 74.0°), 0.484810 (d = 1.172, θ = 61.1°)

our good friend: 0.374607 (d = 1.250, θ = 70.0°), -0.984808 (d = 1.103, θ = 56.0°)

oval shape, an: 0.719340 (d = 1.273, θ = 67.3°), -0.500000 (d = 1.190, θ = 72.8°)

Papyrus, The Ahmes: 0.309017 to 0.358368, 0.906308 (d = 0.958, θ = 61.4°)

Parallax Rung, The: 0.469472 (d = 0.544, θ = 46.6°)

pentagon dodecahedron, regular: 0.104528 to 0.104528

pentagons and hexagons: 0.766044 (d = 1.336, θ = 74.2°), 0.656059 (d = 1.107, θ = 73.9°), -0.992546 (d = 1.002, θ = 30.4°)

photo, takes a: 0.898794 to 0.898794, -0.406737 (d = 1.459, θ = 68.8°), -0.965926 (d = 1.258, θ = 68.9°)

physical model, a: 0.052336 (d = 1.228, θ = 73.0°), -0.017452 (d = 0.700, θ = 69.6°)

pigs, above the: -0.342020 (d = 0.578, θ = 38.5°), -0.974370 to -0.978148

Planetary Defense Mission: 0.848048 (d = 1.470, θ = 64.1°)

Platonic solids, the: 0.275637 to 0.275637, 0.241922 to 0.190809

pocket, into a: 0.669131 to 0.681998

polyhedron shape, biggest: -0.358368 (d = 1.479, θ = 63.4°), -0.629320
    (d = 0.289, θ = 57.4°)
pool hall, a: 0.629320 (d = 0.820, θ = 26.0°), 0.669131 (d = 0.486,
    θ = 41.6°)
principal triangulation, the: -0.857167 to -0.866025
prisms and antiprisms: 0.207912 to 0.190809, 0.156434
    (d = 1.022, θ = 59.2°)
Pythagorean Theorem but: 0.970296 (d = 0.643, θ = 74.3°), 0.838671
    (d = 1.139, θ = 56.7°), -0.559193 (d = 0.701, θ = 27.4°)
Pythagorean Theorem, the: 0.933580 (d = 1.202, θ = 77.7°), 0.965926 to
    0.970296, -0.325568 (d = 1.242, θ = 55.7°), -0.544639 to -0.573576,
    -0.939693 to -0.945519

quad mesh, a: 0.965926 to 0.965926, 0.956305 (d = 0.615, θ = 38.0°)
Queen Mary University: 0.798636 (d = 1.446, θ = 62.8°)
Quite the opposite: 0.422618 (d = 0.596, θ = 46.7°), 0.829038
    (d = 1.230, θ = 80.1°)

rainbow, the second: 0.766044 to 0.777146
rainbows, looking at: -0.469472 (d = 1.258, θ = 74.4°), -0.707107
    (d = 0.914, θ = 57.0°)
raise the question: -0.798636 (d = 0.421, θ = 58.6°), -0.848048
    (d = 1.218, θ = 84.2°)
ran his code: 0.087156 (d = 1.111, θ = 81.9°), -0.104528
    (d = 1.311, θ = 72.2°)
ratio between, the: -0.275637 (d = 0.393, θ = 55.1°), -0.766044
    (d = 0.750, θ = 76.6°), -0.945519 (d = 1.450, θ = 59.4°), -0.933580
    (d = 1.535, θ = 56.9°)
Redshift Rung, The: 0.469472 (d = 1.480, θ = 75.3°)
repeating circle, The: -0.788011 (d = 1.223, θ = 51.1°), -0.848048
    (d = 0.530, θ = 48.5°)
restorative force, a: -0.913545 to -0.913545
Rhombic Dodecahedron versus: 0.642788 (d = 0.415, θ = 72.7°),
    0.587785 (d = 1.166, θ = 58.1°)
rhombic dodecahedron, a: 0.615661 (d = 1.112, θ = 65.0°), 0.587785
    to 0.573576
right angle, a: -0.573576 to -0.573576, -0.694658 (d = 0.636, θ = 43.0°)

skateboard, and the: 0.629320 to 0.629320

SNES, Entertainment System: -0.656059 (d = 0.428, θ = 57.2°)

snub cube, the: 0.241922 (d = 0.803, θ = 26.6°), 0.017452 to 0.000000

snugly together, fit: 0.984808 (d = 0.394, θ = 65.8°), 0.309017
    (d = 1.183, θ = 45.3°)

Space Climate Observatory: 0.173648 (d = 0.786, θ = 36.8°)

Sphere Entertainment Co: -0.190809 (d = 0.755, θ = 44.2°)

sphere, a small: 0.857167 (d = 0.449, θ = 33.1°), -0.052336
    (d = 0.576, θ = 54.7°)

spreadsheet, into a: 0.987688 (d = 0.645, θ = 33.9°), -0.956305 (d = 0.727,
    θ = 72.5°)

square root, take: 0.927184 to 0.927184, 0.956305 (d = 0.879, θ = 73.0°),
    0.970296 to 0.974370, -0.069756 (d = 0.672, θ = 49.1°), -0.374607
    (d = 1.060, θ = 61.7°), -0.719340 (d = 1.359, θ = 76.7°)

stacked and squashed: 0.629320 to 0.629320

Standard Triangle Language: 0.987688 (d = 1.374, θ = 69.3°)

Stane Street They: 0.052336 (d = 0.750, θ = 31.9°)

STL file, an: 0.987688 to 0.987688, 0.978148 (d = 0.896, θ = 18.6°)

street sign, the: -0.990268 to -0.990268

Sun, and the: 0.515038 to 0.515038

Sun, from the: 0.500000 (d = 0.556, θ = 30.9°), 0.731354 to 0.743145,
    -0.731354 (d = 0.955, θ = 55.2°), -0.961262 to -0.961262

Surrey Hills My: -0.207912 (d = 1.060, θ = 43.5°)

Sydler solid, the: -0.034899 to -0.034899

telescope, point a: 0.422618 (d = 1.258, θ = 82.9°), -0.777146 to -0.777146

tell you what: -0.422618 (d = 0.426, θ = 68.5°), -0.573576
    (d = 1.102, θ = 49.7°)

tells us something: 0.601815 to 0.601815, -0.707107 (d = 1.252,
    θ = 69.7°)

tetrahedron, the regular: 0.656059 to 0.656059

The Big Feastival: 0.052336 to 0.052336

The Dice Lab: 0.998630 (d = 1.280, θ = 56.2°)

Thomson problem, the: -0.069756 to -0.069756

three, a mere: 0.999391 (d = 1.359, θ = 76.7°), 0.731354 (d = 0.939, θ = 25.0°)

tiles and paving: 0.777146 (d = 1.127, θ = 61.3°), 0.469472
    (d = 1.255, θ = 84.4°)

why engineers love: 0.981627 to 0.984808
Wile E Coyote: -0.275637 (d = 1.127, θ = 61.3°)
World Trade Center: 0.374607 to 0.374607
wrote some code: -0.406737 (d = 0.923, θ = 82.3°), -0.984808
   (d = 1.025, θ = 59.0°)

X-ray Analysis, An: -0.754710 (d = 0.405, θ = 52.9°)